WORLDS OF THE EAST INDIA COMPANY   Volume 20

AN INNOVATIVE PHYSICIAN AND SCIENTIST IN BRITAIN AND BRITISH INDIA

WORLDS OF THE EAST INDIA COMPANY

ISSN 1752-5667

Series Editor
John McAleer (University of Southampton)

This series offers high-quality studies of the East India Company, drawn from across a broad chronological, geographical and thematic range. The rich history of the Company has long been of interest to those who engage in the study of Britain's commercial, imperial, maritime and military past, but in recent years it has also attracted considerable attention from those who explore art, cultural and social themes within an historical context. The series will thus provide a forum for scholars from different disciplinary backgrounds, and for those whose have interests in the history of Britain (London and the regions), India, China, Indonesia, as well as the seas and oceans.

The series welcomes submissions from both established scholars and those beginning their career; monographs are particularly encouraged but volumes of essays will also be considered. All submissions will receive rapid, informed attention. They should be sent in the first instance to:

Dr John McAleer, Department of History, University of Southampton, j.mcaleer@soton.ac.uk

Other books in the series can be viewed on our website at
https://boydellandbrewer.com

AN INNOVATIVE PHYSICIAN AND SCIENTIST IN BRITAIN AND BRITISH INDIA

THE LIFE AND TIMES OF
SIR WILLIAM BROOKE O'SHAUGHNESSY,
1808–1889

Neil MacGillivray

THE BOYDELL PRESS

© Neil MacGillivray 2025

All Rights Reserved. Except as permitted under current legislation
no part of this work may be photocopied, stored in a retrieval system,
published, performed in public, adapted, broadcast,
transmitted, recorded or reproduced in any form or by any means,
without the prior permission of the copyright owner

The right of Neil MacGillivray
to be identified as the author of this work has been asserted in accordance with
sections 77 and 78 of the Copyright, Designs and Patents Act 1988

First published 2025
The Boydell Press, Woodbridge

ISBN 978 1 83765 191 7

The Boydell Press is an imprint of Boydell & Brewer Ltd
PO Box 9, Woodbridge, Suffolk IP12 3DF, UK
and of Boydell & Brewer Inc.
668 Mt Hope Avenue, Rochester, NY 14620-2731, USA
website: www.boydellandbrewer.com

A catalogue record for this book is available
from the British Library

The publisher has no responsibility for the continued existence or accuracy
of URLs for external or third-party internet websites referred to in this book,
and does not guarantee that any content on such websites is, or will remain,
accurate or appropriate

CONTENTS

Preface		vii
List of Abbreviations		xiii
	Introduction	1
1	From Limerick to Dublin, Edinburgh and London	7
2	O'Shaughnessy and Cholera, Intravenous Saline and Latta	23
3	*Bengal Dispensatory* and *Cannabis Indica*	57
4	Medical Furlough in London and the Royal Society	87
5	'That Man O'Shaughnessy' and Electric Telegraphy	103
	Conclusion	137
	Epilogue	147
Bibliography		149
Index		161

PREFACE

Sir William Brooke O'Shaughnessy was a nineteenth-century Irish-born physician who made remarkable scientific advances in three fields: at the age of twenty-three he explained the pathophysiology of cholera showing that victims died from catastrophic loss of body fluids; his advice as to treatment was unambiguous but for the time almost shocking – that was to replace the lost fluid intravenously with saline. However, this was not his only innovation, for later in India, he carried out drug trials showing that *Cannabis indica* was a useful medication and subsequently introduced it to the West. He then turned his attention to electric telegraphy eventually becoming superintendent of telegraphs for India.

The inspiration to write a book about the life and work of Sir William Brooke O'Shaughnessy took shape during periods of Covid lockdown when it became evident that there was enough material on his life and work to embark on such an enterprise. Sufficient or almost sufficient research information is a *sine qua non* for a biography but more than that is needed to press on with such a project: catalysts are necessary, appropriately so for a book about a man for whom chemistry was the pre-eminent science, almost the only one in his era, and who excelled in chemical analysis. In many ways, there is a peculiar symmetry in starting to write such a biography during an epidemic of a new and frightening disease which came from abroad, from the East, about a man whose first moments of fame occurred during an epidemic of a new and frightening disease which came from abroad, from the East: cholera.

Material on the first cholera epidemic of 1831–1832 and on O'Shaughnessy's role in it arrived almost by accident. When carrying out research on the Edinburgh 1848–1849 cholera epidemic for my doctoral thesis, the discovery of the splendid leatherbound records of Edinburgh cholera patients, in three volumes, maintained in the archives of the Royal College of Physicians of Edinburgh, revealed that some victims were given intravenous saline, albeit in extremely small quantities. Others of course were bled. Intrigued by this, further research led to Professor Robert J. Morris's work on the first cholera epidemic, *Cholera 1832. The Social Response to an Epidemic,* in which he described how O'Shaughnessy had gone to Sunderland, the first English port of call of cholera, to analyse the blood of cholera victims and by careful chemical analysis reached rational conclusions as to why victims died. His conclusions and extraordinary (for the time) recommendations were ground-breaking. Morris went on to relate how Dr Thomas Aitcheson Latta, a young physician practising in Leith, the port of Edinburgh and then a separate burgh, applied O'Shaughnessy's suggested treatment which in effect was to inject large amounts of saline into the circulation. Latta published his results in *The Lancet,* triggering enormous interest and varied reactions. Professor Bob Morris, whose book was so influential in many ways, sadly died recently – he was always encouraging to a novice historian and happy to discuss the events of 1831–1832.

There were other motivating factors: when giving a talk some years ago to colleagues, all Edinburgh medical graduates, on Latta and his remarkably bold use of intravenous saline in Leith and Edinburgh, it became apparent that none of them had heard of him or of his medical world 'first', nor of course had they any knowledge of O'Shaughnessy. This seemed extraordinary – that in the city and medical school where both men had studied they were unknown. Articles have been written on both men in medical and historical journals at intervals down the years but seemingly with little widespread or lasting impact. Every two or three decades there is a new flicker of interest in the first use of intravenous therapy but quickly both O'Shaughnessy and Latta are forgotten.

These intermittent flickers of interest have done little to preserve O'Shaughnessy's name or memory, an omission that this book will attempt to change. O'Shaughnessy made remarkable scientific advances in three distinct fields, but it is his 1831 analysis of the blood and excreta of cholera victims that should be acknowledged as a major innovation. When he recommended intravenous replacement of the body fluids lost in cholera he was flying in the face of orthodox thought. The medical profession in 1831 had no answer to the problems posed by cholera largely because their doctrines were out of date and inadequate, still fixated on humoral theory. Galenic principles were totally inadequate and frequently had a negative impact as exemplified by the obsession with blood-letting. It took several decades before intravenous saline became the standard treatment not only for cholera but for all conditions requiring fluid replacement and for this the name of O'Shaughnessy should be honoured. Nalin, who with a colleague successfully introduced oral replacement therapy, wrote in a far-reaching history of intravenous and oral rehydration of the efforts of Dr William Stevens in England, and of Hermann and Jaehnichen in Moscow but stated that 'the more accurate analyses, pathophysiologic and therapeutic paradigms of O'Shaughnessy inspired the daring clinical applica-tions of those findings of Latta who advanced rational treatment as far as possible in the absence of sterile solutions and a valid therapeutic method'.[1]

---

[1]   David R. Nalin, 'The History of Intravenous and Oral Rehydration and Maintenance Therapy of Cholera and Non-Cholera Dehydrating Diarrheas: A

PREFACE IX

It might have been thought that his departure for India in 1833 would have dimmed his brilliance but in fact he was appointed Professor of Chemistry in the new Calcutta Medical College and was soon involved in research into native medicines, eventually focusing on *Cannabis indica*. He carried out meticulous drug assessment trials, possibly the first such detailed drug trial anywhere – at the same time editing the *Bengal Dispensatory* more or less singlehandedly. On completion of his evaluation, and satisfied of its safety and potential, he recommended its use in specific disorders such as tetanus but also for pain relief, introducing cannabis to the West where it remained a useful medication for well over one hundred years.

O'Shaughnessy was elected as a Fellow of the Royal Society in 1843, proposed and supported by men of distinction in the fields of science and medicine, including the brilliant polymath, Sir John Herschel, mathematician, astronomer and chemist, who was prominent in his support, perhaps seeing in the physician a fellow polymath, a man to be encouraged. Certainly, Herschel's letters to Fellows in the Royal Society library advocating his election express his admiration for his young colleague.

By the time of his election to the Royal Society, O'Shaughnessy had already completed an experimental long telegraph line in Calcutta in 1839 which functioned perfectly and when Lord Dalhousie became governor-general, he appointed O'Shaughnessy in 1852 as superintendent of Indian telegraphs. Within three years over three thousand miles were completed in difficult uncharted terrain. It was for this achievement that he was knighted by Queen Victoria in 1856. Samuel Morse, at a London dinner in his honour in 1856, expressed his admiration for O'Shaughnessy praising his work in India, a tribute from one of the great men of telegraphy. How important the telegraph was in the Uprising of 1857 which threatened British control of the colony is debateable but there is no doubt that thereafter its existence was of great help to the British in the administration of India.

Despite every good intention about writing on O'Shaughnessy, when the time came to press on it was fortunate that another destroyer of inertia appeared: a catalyst. In this case, it involved two persons called Margaret: first Maggie Macdonald, late Archivist to the Clan Donald Archive on the Isle of Skye, a vast clan archive dating from medieval times to the present day, a fruitful source of material on island history, another of my historical interests. Maggie had at one time worked as an archivist at the India Office Records, now maintained in the British Library and was still in contact with a one-time colleague, the second Margaret: Dr Margaret Makepeace, now lead curator of East India Company Records at the British Library and herself the

---

Deconstruction of Translation Medicine: From Bench to Bedside?, *Tropical Medicine and Infectious Disease*, 7, 50 (2022), 1–28, 3, www.mdpi.com/journal/tropicalmed/ (accessed 8 July 2024).

X PREFACE

author of a book on the *East India Company London Workforce*.[2] A meeting with Dr Makepeace in the British Library was extraordinarily useful and led to an introduction to Professor John McAleer of Southampton University and a Zoom conversation with him about Indian history and the East India Company, in particular. This led to an introduction to my editor.

The third catalyst of course is medicine and the Edinburgh Medical School. The New Town O'Shaughnessy knew was more or less complete when he arrived in 1827 from Ireland as a young man of eighteen. The Edinburgh Old Town on the castle ridge close to where the university, the infirmary and the College of Surgeons lay, was still in the process of change but the classical New Town, on the edge of which O'Shaughnessy lived when first in the city, was complete by the later 1820s. The Edinburgh I grew to know coming from the Scottish Highlands in the late 1950s was a city he would have had little difficulty in recognising and is one I remember with great nostalgia. The traditions of a medical school such as that of the Edinburgh Medical School are part of its mystique, but a city's feel and its streets, streets he knew, the route he walked from his 'digs' at the top of Leith Walk to Dr Thomas Hope's lectures in grand Moray Place, passing the house in Heriot Row where twenty years later Robert Louis Stevenson lived, these are all also part of memories. The street O'Shaughnessy lived in, and the very house where he lodged were demolished only in the 1960s.

The fourth catalyst strangely is found in the Isle of Skye where in my crofting township, Camuscross, there were two families in the late eighteenth and early nineteenth century who were involved with the East India Company. The first family includes a kinsman: in the late eighteenth century, John Macinnes, born in Camuscross in 1779, joined the East India Company army in 1798, eventually rising to the rank of general, having over time commanded several regiments of Native Infantry. The author's family croft is in Camuscross, his father, John Macinnes MacGillivray was born in Camuscross, his paternal grandmother born in Camuscross in 1853 was Ann Macinnes, granddaughter of Donald Macinnes, General John Macinnes's uncle. For information on General Macinnes's career in the army of the East India Company, I am grateful to Dr Makepeace.

A second Skye connection with India, although unrelated, involved the Elder family, who established in the late 1790s the herring fishing and curing industry based on the port of Isleornsay, at that time part of Camuscross. Major General Sir George Elder (d. Madras 1836), born in Ross or Inverness, date uncertain, was part of this family. By the middle of the nineteenth century, Colin Elder, nephew of Sir George, had two sons involved in East India

---

[2] Margaret Makepeace, *East India Company London Workers: Management of the Warehouse Labourers 1800–1858* (Woodbridge, 2010).

PREFACE

Company affairs: Major Alexander Macdonald Elder in the army and Benjamin Elder in the Company navy.[3]

I hope that my readers will forgive the author for including personal details. I was born in Gairloch, Wester Ross, and after school there and at Dingwall Academy, studied medicine at the University of Edinburgh. Thereafter, after time in general practice and sailing round the world as a ship's surgeon, I worked in hospitals in Edinburgh, Glasgow, London and Manchester, becoming a Fellow of the Royal College of Surgeons of Edinburgh. I became a consultant otolaryngologist in Blackpool and Lytham St Annes, retiring to study history in Edinburgh where, after an MSc, I took a PhD and I have continued to study, write and publish on medical and Highland history.

There are many scholars and friends who have helped and encouraged a retired surgeon in his endeavours to become an historian. First and foremost: Professor Ewen Cameron, Sir William Fraser Professor of Scottish History and Palaeography, University of Edinburgh who was my PhD supervisor and whose sage advice and friendship continue to inspire and encourage me in my research and writing. During my studies, I was fortunate to meet the late Professor Bob Morris who was an inspiration, discussing cholera 1831–1832, about which he wrote a marvellous book: *Cholera 1832. The Social Response to an Epidemic.*

My colleagues in the Edinburgh History of Medicine Group: Professor Roger Davidson, Professor Gayle Davis, Professor Steve Sturdy, Dr Alan Beveridge, and Dawn Kemp were always a source of ideas and post lecture dinners were both informative as well as great fun. The Scottish Society of the History of Medicine and the British Society for the History of Medicine have both proved invaluable, formally through lectures and conferences and informally at the social events which followed. Another distinguished society, the History Society of the Royal Society of Medicine on whose Council I have served having first been introduced by Dr Anjna Harrar, a medical history scholar; Anjna whose family background is India has always encouraged me in my Indian research. Dr George Venters, an eminent public health physician, born in Newhaven, as was Dr Thomas Latta, has been of great help in establishing Latta's family background and his knowledge of Leith has been invaluable; his article on Latta in *Hektoen International, a Journal of Medical Humanities* is fascinating.

Iain Milne, lately Librarian and Archivist at the Royal College of Physicians of Edinburgh and a member of the Edinburgh history of medicine group, has an unrivalled knowledge of Scottish medical history which he has shared generously and introduced me to many of the College's historical literary gems. The Librarian of the Royal College of Surgeons of Edinburgh has always

---

3   Neil MacGillivray, 'The Congested Districts Board and the Isleornsay Pier, Isle of Skye, 1899–1906', *Northern Scotland*, 13, 1 (May 2022), 45–62.

been helpful in providing details of O'Shaughnessy and Latta who were both licentiates of the College. Many societies and archives have been of great assistance in allowing access to their material: the Royal Society, the Royal Asiatic Society, the National Library of Scotland, the National Records of Scotland, the British Library, the University of Edinburgh Centre for Research Collections, Trinity College Dublin and of course the wonderful Wellcome Collection. Finally, the amazing research carried out on O'Shaughnessy by Séamus O'Donoghue, the founder of Ancestral Line, a genealogical Irish family research online, has been of great help: https://ancestralline.com/Sir-William-O'Shaughnessy-Brooke.php.

I am greatly indebted to my editor at Boydell and Brewer, Peter Sowden, who has guided me through the tortuous process of writing and publishing with tact and care.

ABBREVIATIONS

BL	British Library
CRC	Centre for Research Collections, University of Edinburgh
NLS	National Library of Scotland
NRS	National Records of Scotland
RAS	Royal Asiatic Society
RCPE	Royal College of Physicians of Edinburgh
RS	Royal Society
TCD	Trinity College Dublin
WC	Wellcome Collection

INTRODUCTION

Sir William Brooke O'Shaughnessy (1808–1889), later Sir William O'Shaughnessy Brooke, was a surgeon in the Bengal medical service of the British East India Company from 1833 until his retirement in 1861. This brief description hardly does justice to the career of a man whose innovations in many areas of medicine and science from cholera to telegraphy are the focus of this work.

There have been written brief memoirs and scholarly articles, many of which have concentrated on specific aspects of his life and work and here mention must be made of the late Mel Gorman, Professor of Chemistry at the University of San Francisco, who in *Ambix,* described O'Shaughnessy as a pioneer chemical educator in India; later in *Technology and Culture, The International Quarterly of the Society for the History of Technology* he focused on O'Shaughnessy's role in the establishment of the telegraph system in India. Professor Davis Coakley in *Irish Masters of Medicine* rightly included O'Shaughnessy in the pantheon of Irish physicians and surgeons in his biographical essays recalling the lives of great Irish doctors.[1] However, apart from a short account of his career in India by Adams in 1889 and a more recent article in the *Journal of Medical Biography* no detailed chronicle of his work has hitherto been written.[2] Perhaps his versatility in so many

---

[1]   Davis Coakley, *Irish Masters of Medicine* (Dublin, 1992). Coakley was himself a distinguished physician and historian.

[2]   Mel Gorman, 'Sir William Brooke O'Shaughnessy, Pioneer Chemical Educator in India', *Ambix*, xvi (1969), 107–104; Mel Gorman, 'Sir William O'Shaughnessy, Lord Dalhousie, and the Establishment of the Telegraph System in India', *Technology and Culture, The International Quarterly of the Society for the History of Technology*, 12, 1 (1971), 581–601; J.A. Bridge, 'Sir William Brooke O'Shaughnessy, MD, FRS, FRCS, FSA: A Biographical Appreciation by an Electrical Engineer', *Notes and Records of the Royal Society of London,* 52, 1 (1998), 103–120, in which the author uses O'Shaughnessy's Certificate of Candidature for Election to the Fellowship of the Royal Society to review his many attributes; Neil MacGillivray, 'Sir William Brooke O'Shaughnessy (1808–1889), MD, FRS, LRCS Ed: Chemical Pathologist, Pharmacologist and Pioneer in Electric Telegraphy', *Journal of Medical Biography*, 25, 3 (2015), 186–196.

disparate fields has resulted in a focus by scholars on specific areas in which he excelled rather than a more detailed appreciation of this extraordinary man's innovations.

This study will therefore concentrate on the career in Britain and in British India of this remarkable Irish-born scientist, physician and surgeon. His toxicology research, published in *The Lancet*, before he left for India was remarkable for a young man less than two years out of medical school, but it is his original work on the pathophysiology of cholera during the 1831–1832 epidemic carried out two years after qualifying in medicine in Edinburgh that deserves to be acknowledged as a medical milestone.

Although O'Shaughnessy's professional activities in Britain were extremely brief, lasting only from 1829 to 1833, they are of major interest and significance centred as they are around the cholera epidemic of 1831–1832, the first such to appear in Britain and the cause of official alarm and public terror. His precise chemical analysis of the blood and excreta of cholera victims in Sunderland and Newcastle led to the adoption, albeit briefly, by a Leith and Edinburgh physician, Dr Thomas Aitchison Latta, MD, of intravenous injection of saline as the only treatment likely to save lives. O'Shaughnessy's science-based findings and recommendations and Latta's use of saline intravenously had been forgotten or largely ignored when cholera returned to Britain in 1848–1849 as research has shown.

It was in the early years of the twentieth century before rehydration for hypovolaemia became the treatment of choice and remained so until the introduction of oral rehydration by Nalin and Cash in 1967. O'Shaughnessy's use of science, in particular chemistry, in the investigation of disease was for the time a major innovation and unquestionably stemmed from his study of chemistry and forensic medicine in the Edinburgh medical school, an aspect of his career which will be addressed in the first chapter. As Helen Andrews suggests in her biography in the *Dictionary of Irish Biography*:

> Apart from pioneering a revolutionary life-saving remedy, O'Shaughnessy's originality lay in his prescription of appropriate treatment based on scientific observation and analysis, and also the promotion of chemical analysis as potentially useful in understanding physiological processes and disease. Interest in the treatment waned after the epidemic subsided – Latta died (1833) and O'Shaughnessy left for India – but his contribution was recognised in his election as a FRS (1843), and his work remains the basis of rehydration therapy.

O'Shaughnessy studied medicine first in Trinity College Dublin and then in Edinburgh, taking his doctorate in 1829. He spent about a year teaching practical chemistry in the extramural school and working as an assistant to Professor William Alison before moving to London. There is no obvious

INTRODUCTION

reason as to why he took this step, but it is possible that he became disillusioned with what might be considered as nepotism in certain appointments to the Infirmary and the College of Physicians. Moreover, there were at least three other men teaching practical chemistry in Edinburgh at the time with fierce competition for students. As an outsider he may have found that he was unable to attract enough students and now married making ends meet was not easy. Sadly, he was to find out rapidly that employment prospects might have been problematic in Edinburgh, but he was soon to discover they were even worse in London. He found very quickly that the restrictions on practice imposed by the London Royal Colleges meant that he was totally unable to practise as a physician. The fact that he was an Edinburgh MD with a licentiate from the Royal College of Surgeons of Edinburgh had no standing in the metropolis. He had no wish to practise as a surgeon so this left him with one option: to set up as a forensic analyst where he could make use of his skill in chemistry and forensic medicine, a role that fitted his interests and expertise perfectly. However, undaunted, when the chair of forensic medicine in Kings College, came up he applied for the post but was unsuccessful.

Thomas Wakley, the owner and editor of *The Lancet*, had encouraged the young man publishing his early efforts in toxicology research and into the scandal of food adulteration, a connection that saw him become involved with a pressure group intent on breaking the monopoly and exclusivity of the London Royal Colleges by founding an independent London College of Medicine. O'Shaughnessy was elected secretary of this group, a position which brought him into connect with men of influence, one such being Joseph Hume MP who was chair of the group, radical in politics and who had served as a surgeon with the East Company Medical Service, returning to England as a wealthy man.

When the first cholera epidemic arrived in England, in 1831 he must have been the obvious choice to carry out analyses on its victims but there is a mystery as to who asked him to take on this mission. Whether it was a Vice-President of the London College of Surgeons as he himself later claimed cannot be substantiated but it would seem more likely that it was his colleague Wakley who knew of his abilities. In November 1831, he went north to Sunderland and Newcastle to examine the pathophysiology of the disease by chemical analysis. His remarkable conclusions and recommendations, aided by Thomas Latta, remain as a milestone in medicine although seldom remembered as such, unheralded in Edinburgh or Leith, the city and university in which they had studied.

It is impossible to say how successful or otherwise as a chemical analyst he was in the year after his cholera findings, but finding London's restrictions impossible, he joined the East India Company medical service, specifically the Bengal medical service, one of the three Indian medical services, on 8

August 1833, sailing two days later from London for Bengal with his wife and daughter, arriving in Calcutta in mid-December.[3] His career in Bengal will be reviewed in later chapters, focusing first on further innovative work in the field of pharmacology and therapeutics and his role in the newly established Calcutta Medical College to which he was appointed as the first professor of chemistry in August 1835. During his tenure of the chair of chemistry in the College his research into the therapeutic use of cannabis, and his almost single-handed compilation of the *Bengal Dispensatory and Pharmacopeia* (1841–1844) reveal a scientific mind of enormous ability and perseverance. His later work in experimenting with and then constructing a functioning telegraph system in 1839 in Calcutta was outstanding and in due course led to his appointment in 1852 as superintendent of telegraphs in India, a post which showed his organisational and innovative genius in a discipline remote from chemistry or medicine.

A period of ill-health perhaps precipitated by the strains of his academic position and his well-nigh single-handed editing of the *Bengal Pharmacopoeia* necessitated a spell on medical furlough in Britain in 1841. His recovery on the long voyage to London and later at home saw him sufficiently recovered to travel to Canada as personal physician to Sir Charles Metcalfe (1785–1846) on the latter's appointment as Governor-General. Remarkably, when in Canada he took the opportunity to visit America to meet Joseph Henry and Samuel Morse, two American pioneers of telegraphy. While in London, immediately before leaving for Canada, he was proposed for the Fellowship of the Royal Society in January 1843, his proposers including men of great distinction in the world of science and medicine. His election was a remarkable honour – his proposal and his sponsors will be discussed in a later chapter.

Before he returned to India, he appeared uncertain as to his future writing to his friend Horace Hayman Wilson about his apprehensions, telling that his post as professor of chemistry was no longer vacant. However, on arriving back in Calcutta in 1844 he was appointed assistant master of the Bengal mint and enabled to continue his role in forensic medicine, as advisor to the Bengal government. Three years later, his career took another turn when Lord Dalhousie was appointed governor-general in 1848 and recognised O'Shaughnessy's expertise in telegraphy by appointing him superintendent of telegraphs for India in 1852.

It is interesting to reflect that O'Shaughnessy was born less than a decade after Ireland had been absorbed into the United Kingdom of Great Britain and Ireland, an event that saw the abolition of the Irish Parliament in 1801 and the continued unbroken supremacy of the Protestant Anglo-Irish Ascendancy. In

---

[3]   Dirom Grey Crawford, 'Notes on the Indian Medical Service', *Indian Medical Gazette*, 36 (1901), 3. O'Shaughnessy sailed on 10 August 1833 with his wife and child on the ship *Catherine* under the owner and skipper, Captain Fenn.

INTRODUCTION

the mid to late 1840s, as he was becoming firmly established in the service of the East India Company, back in his native Ireland the Great Famine was causing death on an unimaginable scale as the British state adopted what can only be described as a laissez-faire attitude. There is no record as to O'Shaughnessy's thoughts on the disaster affecting his people back in Ireland nor on the abolition of the Irish Parliament and the incorporating union that resulted. It is possible that he had divided loyalties: Protestant on his mother's side and Catholic on his father's side, aspects of his background that will be examined in a later chapter.

O'Shaughnessy's appointment to the East India Company and his arrival in Calcutta in December 1833 came at a critical time in Anglo-Indian relations; the struggle for language and educational supremacy between the Orientalists and the Anglicists was about to be resolved in favour of the latter with the appearance in Bengal of Charles E. Trevelyan (1807–1886), later Sir Charles Trevelyan, 1st Bart, and, critically, Thomas Babington Macaulay (1800–1859), later 1st Baron Macaulay, who in 1834 became the first law member of the Governor-General's Council. Macaulay's 1835 *Minute on Indian Education* effectively ended the debate in favour of English as the medium of education in Indian secondary schools. Macaulay's Minute argued vigorously for English, 'on the grounds previously advanced by James Mill, that instruction in English would convey the findings of a more advanced culture and so the money would be more usefully spent'.[4] This was a major change in the attitude of the East India Company to native Indian languages and their use in education; the adoption of English as the medium of instruction affected medicine, both in its teaching and practice: the centuries-old Ayurvedic and Unani medicine practised in India was in essence little different, at least in theory, from Western medicine where the Galenic principles of the four humours still held sway and had done so for over 1,500 years. The abolition of the native schools of medicine less than a decade after their foundation was undoubtedly a blow to those who believed that allowing the different languages and cultures to exist side by side was beneficial to both parties.

The replacement of these native schools by the Calcutta Medical College therefore was not without controversy but gave O'Shaughnessy his opportunity when he was appointed the College's first Professor of Chemistry, a major step for a man whose position as an assistant surgeon in the Bengal medical service was hardly one of influence despite his past scholarly achievements. It was undoubtedly fortunate that about this time the East India Company had issued instructions to economise on drugs which were manufactured in Britain and brought from London at considerable expense. They suggested that Indian

---

4   William Thomas, 'Macaulay, Thomas Babington, Baron Macaulay (1800–1859)', *Oxford Dictionary of National Biography* (Oxford, 2004), https://doi-org.ezproxy.is.ed. ac.uk/10.1093/ref:odnb/17349 (accessed 8 July 2024).

medicines wherever possible should be used, a major ruling which encouraged O'Shaughnessy to develop his interest in native treatments, to his assessment of cannabis and his eventual introduction of a new medicament to the West.

At the same time, he was experimenting with photography, presenting his results to the Asian Society at the same time as Daguerre was showing his photographs to amazed audiences in France. But it was not only photography and pharmacology that was occupying his time: he was developing his own technique of electric telegraphy constructing an experimental line which functioned perfectly. This of course led a decade later to his major role as superintendent of Indian telegraphy, a role that earned him a knighthood and the somewhat doubtful accolade that he was the saviour of India through the telegraph system he had constructed that was claimed to have saved India for the British during the Uprising of 1857.

The contribution of Dr Thomas Aitcheson Latta of Leith, an Edinburgh medical graduate, must not be forgotten in the story of intravenous saline. His decision in 1832 to use saline in the treatment of dying cholera patients brought O'Shaughnessy's recommendations to the notice of the medical profession and here the article by Dr A.H.B. Masson in 1971 must be acknowledged as a milestone in the historiography of cholera treatment.[5]

[5] A.H.B. Masson, 'Latta: Pioneer in Saline Infusion', *British Journal of Anaesthesia*, 43 (1971), 681–685.

1

From Limerick to Dublin, Edinburgh and London

This chapter has three sections: the first section will look at O'Shaughnessy's early life, his birth and family before examining his short undergraduate career in Dublin and his time as a student of medicine in Edinburgh. It was as an important part of the study of medicine in the Edinburgh Medical School that chemistry, both practical and theoretic, gave him the analytical tools which enabled him to analyse the deaths of cholera victims in 1831 and to propose treatment that was rational being based on sound scientific principles. The second part will look at his early publications in *The Lancet*, his friendship with Thomas Wakley, owner and editor of that journal and his truncated career in London where circumstances (perhaps fortunately) obliged him to pursue work as a chemical analyst rather than as a physician.

The chapter that follows will focus on O'Shaughnessy's pathophysiological research and what has been described as his 'valid therapeutic paradigm for correcting the pathophysiological disorder' which led to the implementation of his proposed treatment in 1832 by Dr Thomas Latta in Leith and Edinburgh.[1] O'Shaughnessy proposed intravenous fluid replacement as the rational treatment for the devastating loss of body fluids which was the cause of death in cholera and in time became this became the standard until the advent of oral replacement therapy in the 1960s. This translational medical advance, it will be argued, took medicine and therapeutics from a world still largely dominated by ancient humoral theory into a science-based era where research based on chemical analysis and the application of rational treatment arising from that research were linked.

A scholar has written [that] 'In the whole of the history of therapeutics before the twentieth century there is no more grotesque chapter than that on the treatment of cholera, which was largely a form of benevolent homicide.' He concluded 'medicine has always needed and still needs, more science – not

---

[1] David R. Nalin, 'The History of Intravenous and Oral Rehydration and Maintenance Therapy of Cholera and Non-Cholera Dehydrating Diarrheas: A Deconstruction of Translational Medicine: From Bench to Bedside?, *Tropical Medicine and Infectious Disease*, 7, 50 (2022), 1–28.

8   AN INNOVATIVE PHYSICIAN AND SCIENTIST

more compassion'.[2] It will be argued that O'Shaughnessy used his scientific training for rational ends.

It is interesting to reflect that O'Shaughnessy was born less than a decade after Ireland had been absorbed into what became the United Kingdom of Great Britain and Ireland, an event that saw the abolition of the Irish Parliament in 1801 and the continued unbroken supremacy of the Protestant Anglo-Irish Ascendancy. There is some understandable confusion as to O'Shaughnessy's origins and birthplace: at the time of his birth, it appears that the family name had recently been changed from O'Shaughnessy to Sandes. There is no question as to his paternity: his father was Captain Daniel O'Shaughnessy from Limerick, otherwise Sands or Sandes, who it is alleged took the name Sandes at the request of a maternal uncle by name McMahon who 'had realised a large property in India under the patronage of Sir Philip Francis, which he promised to bequeath to Captain Sandes, and his brother, Dr Sandes, who was an attending physician of the County of Limerick Infirmary'.[3] Whether the McMahon in question was Sir John McMahon Bart (1754–1817) who was born in Limerick, son of John McMahon, comptroller of the port of Limerick, cannot be established with any certainty but the connection with the harbour may be significant in view of Daniel Sandes' involvement with shipping. The timing of the death of Sir John in 1817, who died without issue, the baronetcy passing to his half-brother, Thomas, may also support the possibility. However, there is no satisfactory answer as to why the name Sandes was chosen as opposed to McMahon, the maiden name of Daniel's mother. It begs the question also as to why the name O'Shaughnessy was unacceptable to the uncle, a riddle to which there is no answer.

His mother's ancestry is more straightforward and also reveals family connections with India. William's mother was Sarah Boswell, born in Dublin, the daughter of John Boswell, a merchant and his wife Henrietta Brooke, both members of the Protestant Church of Ireland. Henrietta's first cousin was Lieutenant-General William Joseph Eyre Brooke (1770–1843), who was born in India and who had served in the West Indies and the Peninsular War; he was the son of Henry Brooke (c. 1725–1786) of County Kildare who was an administrator in the Madras Civil Service.[4] Henry Brooke married Mary

---

   [2]   Norman Howard-Jones, 'Cholera Therapy in the Nineteenth Century', *Journal of the History of Medicine* (1972), 374, 393.

   [3]   *Dictionary of Irish Biography* (2009). Sir Philip Francis is dealt with under the name of his father, John Francis.

   [4]   *The Journal of the Kilkenny and South-East of Ireland Archaeological Society*, 6 (1871), 247; Sir Philip Francis (1740–1815) was an Irish politician who was from 1773 a member of the Supreme Council of Bengal at a salary of £10,000 per annum. Disagreements and a duel with Warren Hastings, the governor-general, led to his return to England in 1780.

FROM LIMERICK TO DUBLIN, EDINBURGH AND LONDON

Aubrey, Allbury or Allbeary (c. 1738–1787) at Fort St George, Madras in 1764; it has been alleged by some authorities that Mary was the illegitimate daughter of Prince Frederick of Wales (1707–1751), son of King George II and father of George III. Whether there is any truth in this allegation cannot be ascertained. What is striking is how frequently India and service in that country appear in William's ancestry on both sides of his family and such connections may well have influenced him in his eventual choice of career with the East India Company.

William was baptised a Catholic under the name O'Shaughnessy although his parents' marriage had been solemnised in the Protestant Church of Ireland in County Kerry. There is also some doubt in articles and biographical entries as to the year of his birth, some authorities stating he was born in October 1808 whereas others claim 1809 as the year, including the writer of his entry in the *Dictionary of Irish Biography*, but in this work it will be recorded that 1808 is the correct year, a fact proven by birth and baptism records. His father, Daniel Sandes or O'Shaughnessy, has been described variously as a merchant or shipping agent or broker in the City of Limerick but of course the captain prefix hints at a sailing background. Many years later, in 1870, William's third marriage certificate records his father as a shipowner from Limerick and the earlier reference to Captain Sandes confirm a maritime occupation. Daniel was listed as bankrupt in 1815 and it is possible that his father's bankruptcy was the one of the reasons that William came to be enrolled in the grammar school in Ennis, perhaps relying on funds from his mother's family to pay his school fees and of course as adherents of the Church of Ireland they would favour a Protestant school.

The Ennis grammar school was one of the Erasmus Smith Schools, an educational charity established by Royal Charter in 1669 after its initial foundation under Oliver Cromwell and as a rule only admitted Protestant students, but it seems that Catholic boys were occasionally permitted entry. As a biographer points out:

> The prime purpose was to teach the sons of Smith's Irish tenants 'fear of God and good literature and to speak the English tongue'. As well as grammar and 'original tongues', utilitarian skills such as reading, writing, and accounts were on the curriculum. The most promising pupils were to be further aided by scholarships to Trinity College in Dublin.[5]

According to an article on education in County Clare by J. Power of Clare County Library, Marcus Paterson (1712–1787), Solicitor General and Lord Chief Justice of Ireland, himself an Ennis man, was largely instrumental in

---

[5] T. Barnard, 'Erasmus Smith (bap. 1611, d. 1691), Merchant and Educational Benefactor', *Oxford Dictionary of National Biography* (Oxford, 2004).

bringing this school into existence around 1773.[6] It is likely that his mother's religion and family influence through the Brooke connection were factors in enabling William to go to Ennis grammar school, largely a Protestant foundation.

Unusually for the early nineteenth century, the Erasmus Trust grammar school was not the only school in Ennis at this time – Power's work on education tells us that 'Mr. Cole's English and French Academy at Arthur's Row, Ennis, under the patronage of the bishop of Killaloe, opened in 1807. Young gentlemen were taught English, French, geography, writing and arithmetic in preparation for university or for the Royal Military College at Marlow.' Many scholars have stated that William's family was related to the Revd Dr James O'Shaughnessy (1745–1828), Catholic bishop of Killaloe from 1807 to 1828, who has been in several sources referred to as William's great uncle, but there is no genealogical evidence to prove such a relationship. Nevertheless, the shared surname does indicate at least that there might indeed be a degree of kinship. The Dean of Ennis, Revd Terence O'Shaughnessy, has been named as an uncle but again there is no proof of such a connection. The bishop's obituaries described him as descended from the O'Shaughnessy family of Gort, an ancient aristocratic family whose lands were in the district of Kinelea in County Galway, close to the border with County Clare. Their estate was confiscated and given to Sir Thomas Prendergast, 1st Baronet in the late seventeenth century, the last of the chiefly line being Colonel William O'Shaughnessy who died in France in 1744. The only fact of ancestry and kinship of which we can be sure is that William was descended also from the ancient aristocratic O'Shaughnessy family whose lands lay in Galway and through this descent may have been related to the bishop and the dean.

From the Erasmus Smith grammar school in Ennis, William at the age of seventeen was accepted by Trinity College Dublin to study medicine, matriculating on 17 November 1825 and recorded in the medical school register as William Sands O'Shaughnessy, a Roman Catholic born in Clare. It is noteworthy that as late 1825 the Sands name was still being used by William

---

6   Erasmus Smith Schools is an educational charity established by Royal Charter in 1669 after its initial foundation under Oliver Cromwell. It was known for many years as The Erasmus Smith Trust or as 'The Governors of the Schools Founded by Erasmus Smith, Esquire'; the Ennis Grammar School was endowed by this trust and took mostly boys of a Protestant background, although Catholics were accepted also. The *Oxford Dictionary of National Biography* states that the Trust also funded scholarships at Trinity College Dublin. I am most grateful to Alan Phelan, Archivist to the Erasmus Smith Schools for his help in clarifying matters concerning the school in Ennis.

and the Brooke name was not yet apparent. The Erasmus Smith Trust funded scholarships largely for boys of the Protestant faith at Trinity College, but occasionally Catholic boys were given scholarships which might explain his acceptance by a largely Protestant establishment. Although Trinity College was primarily a Protestant institution, from 1793 Catholics were permitted to enter for study and to take degrees and it appears that William benefited from this freedom. It is also possible that he had an Erasmus Smith scholarship at the college: in the matriculation register he was recorded as a 'Pensioner', a term indicating that his fees were being paid by some person or organisation, not specified, and were fixed at an annual fee. His *Ludimagistri* or teacher was noted as Revd Fitzgerald and the records of the Erasmus Schools confirm that Revd Michael Fitzgerald was the master during the period when William was a pupil there, confirming beyond doubt that O'Shaughnessy was educated in Ennis at the Erasmus Smith grammar school.[7]

O'Shaughnessy, after spending about eighteen months as a student at Trinity College, transferred in 1827 to the University of Edinburgh medical school. His reasons for this are not known but Francis Barker (1773–1859), Professor of Chemistry at Trinity College from 1808/1809, had studied in Edinburgh taking his MD there in 1795 with a thesis entitled: '*De animalium electricite*.' His biographers suggest that he was primarily a physician rather than a chemist, but there is every likelihood that attending the lectures of Joseph Black (1728–1799), Professor of Medicine and Chemistry in Edinburgh and briefly those of Thomas Hope (1766–1844), Black's assistant from 1795 and successor in 1799, influenced Barker, persuading him of the value of chemistry and of Edinburgh's pre-eminence in the teaching of the subject. Such a realisation and recognising the considerable abilities of the young O'Shaughnessy who by this time was no doubt interested and increasingly proficient in chemistry, a relatively new and exciting science, may have encouraged Barker to advise him to study in Edinburgh. There is no surviving record to explain why he took this step, and we can only surmise as to the reason or reasons for this decision.

The deciding factor may simply have been the renown of the Edinburgh medical school. It is generally accepted that in the early decades of the nineteenth century the Edinburgh school of medicine was without equal in Britain and this reputation may have been a factor in his decision. Edinburgh's fame as a medical school was international, as a glance at the matriculation and graduation registers of the period reveals when men from all parts of the world appear. Anderson says that 'around 1830 the university was at a numerical peak' with over 2,100 students in 1825, 800 of them studying medicine. It was the presence of teachers such as Thomas Hope (1766–1844) in chemistry,

---

[7]   I am grateful to Ellen O'Flaherty of Trinity College Dublin for these details from the Medical School Register (TCD MS 758).

Robert Christison (1797–1882) in forensic medicine and William Pulteney Alison (1790–1859) in the institutes of medicine that drew many students to the city.

It was not simply the reputation of the University School alone that attracted students to Edinburgh but also the growth of a separate extramural school of medicine which existed quite independently from the university, functioning in parallel, and classes taken here often counted towards the degree in medicine. Teachers in this unofficial school were often of equal or even greater eminence to those in the university and attracted students in considerable numbers, perhaps the most notable, later infamous, of these being Dr Robert Knox (1791–1862) in anatomy. A student intending to graduate with a medical doctorate was required to study the following subjects: Anatomy, Botany, Chemistry, Surgery, Principles and Practice of Medicine, Physiology and Medical Jurisprudence. Some of these subjects could be studied outside the university in classes given by lecturers from the extramural schools where many were rivals to the university professors in their ability to attract students and their fees and such was the quality of the teaching that they 'constituted a pool of ability for filling chairs and of course attendance at their classes often counted towards a degree'.[8]

When O'Shaughnessy at the age of eighteen came to Edinburgh in May 1827, he signed the matriculation album as William Brooke O'Shaughnessy from Ennis, Ireland with no mention of Sands, the name he had used when he matriculated in Dublin in 1825 and perhaps significantly on this occasion the middle name Brooke was substituted.[9] The university archives have the only surviving record of O'Shaughnessy attending classes and lectures in the university medical school and that is the Botany class of winter 1828–1829 under the Professor of Botany, Robert Graham (1786–1845), who will appear later in the story of cholera. There is no record in individual class lists of other classes he took in either the university medical school or in the extramural school, but prior to his graduation in 1829 he is documented as having taken classes in Clinical Medicine, Botany, Obstetrics and Clinical Surgery. How many of the classes he required to take for his degree were taken in the extramural school is unknown. Gorman, describing O'Shaughnessy's role as a chemical educator in India, reviewed his time as a student in Edinburgh, finding no record of his having signed up for the chemistry classes of any of the university teachers. Nevertheless, there is no question that the study of chemistry and practical chemistry had become of the utmost importance and in O'Shaughnessy's case private extramural classes must have answered the

---

8   Robert D. Anderson, 'The Construction of a Modern University: Age of Reform', in Robert D. Anderson, Michael Lynch and Nicholas Phillipson (eds), *The University of Edinburgh: An Illustrated History* (Edinburgh, 2003), 113–114.

9   University of Edinburgh Medical Matriculation Index 1783–1968.

need for teaching. Anderson comments that 'private classes in chemistry have to be seen in the context of the extramural teaching of medicine as a whole. There can be no doubt that in the early years of the nineteenth century many students opted for private courses.'

It has been stated that Hope's classes 'included medical students for whom his lectures were compulsory', and it is clear therefore that O'Shaughnessy must have attended his lectures without any official record of him doing so surviving. Moreover, he almost certainly attended the classes of one of the many extramural lecturers. In chemistry and practical chemistry there were several notable lecturers both in the university and in the extramural school and these men undoubtedly had a great influence later on O'Shaughnessy and his career.[10] Although there is no record of him attending classes in chemistry or practical chemistry, since it was a stipulation that chemistry was obligatory for a doctorate in medicine and also for a diploma from the College of Surgeons of Edinburgh, the licentiate he took in March 1829, he must have taken classes on both theoretical and practical chemistry. A review of the teaching of chemistry in the university reveals that there were four lecturers in practical chemistry alone during the period when O'Shaughnessy was in Edinburgh, with others lecturing in the extramural school.[11] It is relevant that in 1829 the Royal College of Surgeons of Edinburgh decreed that candidates for its licentiate undertake a three-month practical chemistry course; O'Shaughnessy must therefore have fulfilled the requirements of the college when he became a licentiate in 1829 and had therefore completed a course in practical chemistry.[12]

The later significance of practical chemistry in his career in Edinburgh, London and Calcutta is proof of his interest and ability learned at the classes of men such as David Boswell Reid (1805–1863). In 1827, Reid started to teach practical chemistry in High School Yards, close to the university and the College of Surgeons and had been given a special licence by which students were allowed to use his 'ticket' as part of their qualification for the LRCS Ed. From 1828, he assisted Professor Hope giving instruction in practical chemistry although the relationship was never entirely happy and was to deteriorate over time. Edward Turner (1796–1837), from 1823, taught a course lasting six

---

[10] Jack B. Morell, 'Practical Chemistry in the University of Edinburgh, 1799–1843', *Ambix*, 16 (1969), 68–80, discusses in detail Hope's tenure of the chair of chemistry and the development of practical chemistry classes mainly by extramural lecturers.

[11] Mel Gorman, 'Sir William B. O'Shaughnessy, Pioneer Chemical Educator in India', *Ambix*, 30, 2 (1983), 107.

[12] For a detailed review of the examination regulations of the Royal College of Surgeons of Edinburgh, see Helen Dingwall, 'Enlightenment to Reform, Incorporation to College 1726–1830', in *A Famous and Flourishing Society: The History of the Royal College of Surgeons of Edinburgh, 1505–2005* (Edinburgh, 2005).

months in chemistry and practical chemistry in the extramural school, until he went to London in 1828; he published *Elements of Chemistry* in 1827 which went through eight editions in ten years, being revised and enlarged after his untimely death by his brother William, aided by Justus von Liebig and William Gregory. Robert Christison considered Turner's book to be the best on the subject in the English language. It is probable that both Turner and Reid had considerable influence on O'Shaughnessy's development as an expert practical chemist.[13] Moreover, in view of his later interests and proficiency and the evidence available, it is certain that he studied forensic medicine (medical jurisprudence), attending classes and lectures given by Robert Christison.

When he matriculated in the university for the session 1826–1827 the register shows that there were 858 men studying medicine, ninety-three of them from Ireland. As generations of students thereafter did, the author included, he signed the Matriculation Register in the University Old College, a building designed by Robert Adam, constructed between 1789 and 1828. It was here in the Old College that the Medical School existed until the New Medical School in Teviot Place was completed in 1888; less than a mile from his lodgings in The Terrace, Leith Street, O'Shaughnessy would have attended lectures in the College and 'walked the wards' in the Royal Infirmary a few dozen yards away at the bottom of Infirmary Street.[14]

The city of Edinburgh in 1827 had undergone enormous changes over the previous fifty years, years which saw the development of the architecturally magnificent New Town to the north of the city, a neighbourhood to which the middle and upper classes had moved, abandoning the increasingly overcrowded, squalid and insanitary medieval city on the ridge between the Castle and the Palace of Holyrood. During the late eighteenth century and the nineteenth century, when aristocrats, lawyers, doctors and other men of substance fled to live in the splendid New Town, the Old Town was now mostly abandoned by the middle and upper classes becoming the home of Edinburgh's poor and increasingly a slum. However, the offices of the city council, the law courts, the infirmary, the College of Surgeons and the University remained, thus the overcrowded tenements of the wynds and alleys of the evermore neglected Old Town would have been a familiar sight to O'Shaughnessy as

---

[13] H. Reid, *Memoir of the Late David Boswell Reid* (Edinburgh, 1863), 8, 9; Robert Christison, *Biographical Sketch of the Late Edward Turner M.D.* (Edinburgh, 1837), 18, 30. Christison considered that Turner's *Elements of Chemistry* was the best text on chemistry in the English language. Turner studied in both Gottingen and Paris at the same time as his friend and colleague, Robert Christison; W.H. Brock, 'Turner, Edward', *Oxford Dictionary of National Biography* (Oxford, 2021); Edward J. Gillin, 'Reid, David Boswell', *Oxford Dictionary of National Biography* (Oxford, 2016).

[14] The Terrace at the top of Leith Street where O'Shaughnessy lodged is no longer in existence having been demolished during redevelopment in the 1960s.

FROM LIMERICK TO DUBLIN, EDINBURGH AND LONDON

he walked from his lodgings in Leith Street to classes in the newly completed Old College or to the nearby Royal Infirmary. This teeming slum, most of whose inhabitants lived in unimaginable squalor, was desperately overcrowded and deteriorated further when fires destroyed several tenement buildings in the 1820s and new road developments destroyed more, leaving thousands of people homeless. The contrast between Old Town squalor and New Town elegance, between disease and plenty, must surely have affected the young medical student.

The Edinburgh extramural school had in Robert Knox (1791–1862), the anatomist, the most renowned of these extramural teachers; Knox later became infamous as a result of his association with Burke and Hare whose murderous procurement of bodies for Knox eventually ruined his reputation. On the retirement in 1824 of John Barclay (1758–1826), who had conducted highly successful anatomy classes at 10 Surgeon's Square, Knox took over his premises and his classes and soon developed an enormous reputation; within four years his anatomy class had over 500 students. Comrie in his two-volume work on the history of medicine in Scotland contends that 'the ineptitude of the third Monro' helped both Barclay and Knox enormously; the third Monro referred to was Alexander Monro (tertius) who succeeded his father Alexander Monro (secundus) in the chair of Anatomy, who in turn had succeeded his father Alexander Monro (primus) in the chair. Knox was a gifted anatomist, easily outshining the reportedly uninspiring teaching of the professor, Alexander Munro Tertius (1773–1859); Charles Darwin (1809–1882), an Edinburgh medical student 1825–1827, was especially dismissive of Munro finding his lectures dull, as did his brother, Erasmus, a medical student in Edinburgh at the same time. Charles was appalled not only at the poor quality of the lectures but also by the unkempt appearance of Professor Munro Tertius.[15]

Knox's reputation was well deserved since there is little doubt that he was one of the foremost teachers of anatomy of his time, but in late 1827, only a few months after O'Shaughnessy's arrival in Edinburgh, Knox's career and reputation was destroyed by his purchase of bodies supplied by Burke and Hare and murdered by the two. The trade in bodies between the two 'body snatchers' and Robert Knox began in November 1827 when Knox from his base in Surgeon's Square accepted a body from them with no questions asked as to its origin. The story of the two Irish-born criminals who had also moved to Edinburgh in 1827, settling in Tanner's Close in the Old Town, is well known and still arouses curiosity. It is thought that in the course of their terrible trade they murdered at least sixteen people, selling the bodies to Knox, each for the

[15] For details of Darwin's time as an Edinburgh medical student, see www.darwin project.ac.uk.

then not inconsiderable sum of £7.10 shillings.[16] When the activities of Burke and Hare were revealed to a shocked city and an equally shocked university, the medical school was exposed as complicit, having turned a blind eye to the activities of grave-robbers and others who provided the medical school with bodies for dissection. There is every likelihood that O'Shaughnessy attended demonstrations and lectures by Knox from first arriving in the city unaware of the criminal activities going on around him. In November 1828, Knox's collusion in their dreadful crimes was revealed. Knox was interviewed by Robert Christison, one of two forensic pathologists appointed to aid the investigation. Whether the fact that the two murderers were from Ireland affected the recently matriculated medical student, also Irish, is impossible to say but the whole episode must have left its mark.[17]

Although little is known in detail about O'Shaughnessy's studies and his life in general in Edinburgh, more is known about where he lodged as an undergraduate and the fact that he married before he graduated is also well documented. He was not yet twenty-one years of age when he married in March 1829 and would not graduate in medicine until July 1829, but he did have a licentiateship from the Royal College of Surgeons of Edinburgh, dated 3 March 1829 which is the reason he was able to describe himself as a surgeon on the marriage certificate. He married Isabella Lawrence in St Cuthbert's Parish Church of Scotland on 30 March 1829, both parties living at number 11, The Terrace, Leith Street; further research shows that at this address lived Helen Lawrence, who in the Edinburgh Post Office Directory for 1828–29, is described as a keeper of lodgings.[18] It appears therefore that O'Shaughnessy married his landlady's daughter. The circumstances surrounding this marriage are unknown, but the fact of the marriage is itself remarkable – because of his age, undoubted religious differences, his wife being presbyterian and perhaps also social class, the latter of considerable importance in an Edinburgh society at that time obsessed with class differences.

Unusually for a doctor after graduating in medicine, he matriculated again for the university year 1829–30, and by 1 February 1830 he had moved away from Leith Street presumably with his wife; in a paper published in *The Lancet* in February of that year his address was given as 65 Lauriston Place, a street relatively close to the university and the infirmary.[19] After graduation, he became Clinical Assistant to Professor William Pulteney Alison (1790–1859), Professor in the Institutes of Medicine, a distinguished physician and

[16] J.D. Comrie, *History of Scottish Medicine* (London, 1932), vol. 2, 500–502.

[17] Owen Dudley Edwards, *Burke and Hare* (Edinburgh, c. 1984). The author, an Irish-born University of Edinburgh historian, recounts the crimes perpetrated by the two men.

[18] *Post Office Directory for Edinburgh* 1828–29, 103.

[19] *Post Office Directory for Edinburgh* 1830–31, 144.

philanthropist, with a strong social conscience who was one of the founders of the Edinburgh New Town Dispensary in 1816.

It is certain from later comments that O'Shaughnessy was by now teaching and demonstrating practical chemical techniques to students in Edinburgh, more especially techniques used in forensic pathology.[20] *The Lancet* article of February 1830 is a remarkable first academic publication from a recent graduate in medicine who at the time of writing was a mere twenty-one years of age. He wrote of how he had discovered errors in recommended tests for the poison, nitric acid: 'having been very recently engaged in a series of toxicological experiments, my attention was particularly arrested by the discovery of a glaring fallacy in one of the tests recommended by the most eminent authorities for the detection of this important poison'.[21] He followed this with a letter to *The Lancet* in which he responded to differences of opinion on the topic pointing out that in Edinburgh he had 'discovered and demonstrated to my pupils certain facts about these tests which had been disputed by two other authors of papers'. This certainly confirms that he was conducting classes in practical chemistry in Edinburgh before he left for London.[22] The 'most eminent authorities' whose results he criticised included Justus von Liebig (1803–1873), considered to be one of the founders of organic chemistry, also his former teacher of forensic pathology, Robert Christison, one of the foremost toxicologists of his day, but he also mentions Edward Turner (1796–1837) and David Boswell Reid (1803–1863), two men with publications to their names, chemistry teachers in Edinburgh, both of them recognised experts in their field. It is an indication of the times that none of these 'eminent authorities' were very much older than O'Shaughnessy himself, two in their twenties and two in their thirties. Gorman recounts how his criticism resulted in a flood of letters to *The Lancet* defending the accuracy of Liebig's nitric acid test, but O'Shaughnessy rebutted them all pointing out that in forensic toxicology where criminality was being suspected there could be no room for doubt.[23]

Despite his position as assistant to Professor Alison, his research work in chemical toxicology and his teaching, it seems that his work in Edinburgh

---

[20] Gorman cites personal information from the Chief Librarian, University of Edinburgh Library regarding O'Shaughnessy's post as clinical assistant to Alison: Mel Gorman, 'Sir William B. O'Shaughnessy, Pioneer Chemical Educator in India', *Ambix*, 30, 2 (1983), 55; Robert Christison, *A Treatise on Poisons in Relation to Medical Jurisprudence, Physiology, and the Practice of Physics* (Edinburgh, 1829).

[21] W.B. O'Shaughnessy, 'On the Mode of Detecting Nitric Acid and the Total Inefficacy of the New Test of Decolorizing the Sulphate of Indigo', *The Lancet*, 14, 352 (29 May 1830), 330–333.

[22] *The Lancet* (June 1832), 302–303.

[23] Gorman, 'Sir William B. O'Shaughnessy, Pioneer Chemical Educator in India', 55.

18 AN INNOVATIVE PHYSICIAN AND SCIENTIST

did not satisfy his professional ambitions or possibly, he recognised the difficulties he might encounter in achieving promotion in Edinburgh. It was at
this juncture that James Craufurd Gregory, not long returned from Paris, was
positioning himself to succeed Christison in the chair of forensic medicine, a
circumstance which will be explored in a later chapter. O'Shaughnessy would
have known only too well the challenge he faced. He was not from Edinburgh;
his religion was not readily accepted in Scotland and his marriage to a woman
possibly not of his social class would have counted against him. Whether the
environment in London would be any more accepting he was soon to find out.

O'Shaughnessy moved to London in 1830, by which time he had published
several papers in *The Lancet*, the ground-breaking medical journal founded in
1823 by Thomas Wakley (1795–1862); both the journal and its proprietor were
to play significant roles in his career before 1831 was over. In 1831, now living
in Camberwell, London, he published a lengthy monograph composed of three
essays which he had translated from the French on the treatment of scrofula by
the use of iodine; scrofula was the term then used for an infection of the lymph
nodes in the neck, tuberculosis being the most common cause. The essays
were the work of Dr Jean G.A. Lugol (1786–1851), a Paris physician, and are
a lengthy exposition of his practice with appendices added by O'Shaughnessy
recording the experiences of several other French physicians. This substantial
translation was not his only occupation at this time it seems – encouraged
by Wakley he began to investigate the very widespread adulteration of food,
especially of confectionary, publishing again in *The Lancet* and writing:

> it is my principal aim to lay before the public and the medical profession a
> calm dispassionate statement of the existence of various poisons (gamboge,
> lead, copper, mercury, and chromate of lead in several articles of confec-
> tionary the preparation of which from their peculiar attraction to the younger
> branchs of the community, has grown into a separate and most extensive
> branch of manufacture.[24]

It became clear to him shortly after his arrival in London that the restrictions
imposed by the London Royal College of Physicians would not allow him to
practise as a physician and since he had no wish to be a surgeon his options
were extremely limited. Despite having a doctorate in medicine from the
University of Edinburgh, then arguably the best school of medicine in Britain
and being a licentiate of the Royal College of Surgeons of Edinburgh, an

---

[24] W.B. O'Shaughnessy, 'Poisoned Confectionary', *The Lancet* 2 (1830–1831),
193–198. This article was part of Wakley's campaign against food adulteration in
association with O'Shaughnessy who carried out the chemical analysis. Earlier, similar
work had been done by the German chemist, Frederick Accum (1769–1838), *A treatise
on adulterations of food and culinary poisons* (London, 1807).

FROM LIMERICK TO DUBLIN, EDINBURGH AND LONDON 19

older institution than its sister London College, these qualifications were not deemed adequate by the London Colleges. It does of course beg the question as to whether or not he knew of these limitations on practice before he moved to London. Perhaps Edinburgh had become intolerable because of blatant nepotism exemplified by the rapid promotion of James Craufurd Gregory.

His frustration and anger at such blatant prejudice in London was of course the reason he joined Thomas Wakley in a group of medical men and others seeking to establish a London College of Medicine, independent of the two London Royal Colleges but with the power to confer degrees and permit practice within the metropolis. That he was speedily appointed as secretary of the group is testimony to his rapid acceptance by at least some of his peers in the city.

The first official meeting of this group took place on 16 March 1831 although informal meetings had taken place earlier. The chairman, Joseph Hume MP, was a Scotsman, qualified in medicine from Edinburgh, who had served from 1799 as a surgeon in the East India Company Bengal medical service. On his return to Britain having made a considerable fortune in India, reputedly 40,000 pounds, he purchased the seat of Weymouth, becoming an MP. He associated with James Mill, a schoolmate from Montrose, and Jeremy Bentham, in the group known as Philosophical Radicals.[25] By the time of the second formal meeting on 4 May 1831, Dr W.B. O'Shaughnessy had been appointed as the honorary secretary of the pressure group. Prior to his reading of the formal report which the committee had prepared 'relative to the permanent establishment and future government of the new College' he made a statement regarding his own position:

> I am a graduate of the University of Edinburgh. Circumstances induced me to remove from that capital to London, and the moment I arrive here, I find myself totally unable to practise my profession. I attempt to practise as a physician; I am met by the sneers and reflections of my fellow practitioners, "that I am not a licentiate of the London College." I am not a surgeon, therefore I do not practise in that department of the profession.

He continued:

> young as I may be, and inexperienced as I may be, I consider myself imperatively called on to come before the public and advocate as warmly as is within my power the cause of medical reform.

---

[25] *The Lancet*, 15, 395 (26 March 1831), 846–886, in which is a verbatim report of Hume's address to the pressure group; V.E. Chancellor, 'Hume, Joseph, (1777–1855)', *Oxford Dictionary of National Biography* (2004); James Mill (1773–1836), a Scottish historian, Greek scholar, graduate of the University of Edinburgh and author of *History of India* (1818), father of the better-known, John Stuart Mill.

This impassioned speech from, as he admitted, a young and inexperienced man, was greeted with 'great cheering'. It is surely a mark not only of his self-confidence but also the regard with which he was held by his London peers that he was appointed as honorary secretary and was able to speak as he did.[26]

Though ultimately unsuccessful, the London College of Medicine pressure group incorporated ideas that formed the basis of reforms in the charters of the main licensing bodies: the Society of Apothecaries and the Royal Colleges of Surgeons and Physicians. The new proposed College was to be a single faculty, therefore including physicians, surgeons and general practitioners; moreover, teachers at private medical schools and naval surgeons would also be included. It was to be free of all religious restrictions, a move aimed at Oxford and Cambridge Universities both of whom required students to belong to the Church of England. Its officers and Senate were to be decided by annual ballot. The cost of diplomas would be set low and those already qualified would be eligible to become Fellows, thus allowing men qualified in Scotland to be admitted without re-examination. Appointments to official (public) positions were to be by merit, eliminating nepotism and the hand-placing of protégées. All Fellows would carry the prefix 'Dr', removing artificial divisions between members.[27]

Perhaps unsurprisingly, the London College of Medicine failed when faced with the united opposition of the established Royal Colleges and the Society of Apothecaries. The imminent arrival of cholera must also have concentrated their minds on other more pressing matters. Nevertheless, a strong case for reform had been made in the most public manner; subsequent legislation and reforms in governing charters were, for many years, influenced by this campaign. The total failure to rationalise the affairs of medical institutions is shown by the fact that by the late twentieth century the number of Colleges in London had multiplied considerably perhaps indicative of a resistance to democratic and unifying measures. On a personal level, O'Shaughnessy had impressed the prime mover, Wakley, in this attempt to bring some form of democracy to London medicine and this would have a bearing on his immediate future. During the next three years, Wakley's intentions to carry on his agitations against the College of Surgeons were not altered but he decided to change his methods. He had done as much as was possible in his private capacity and as a journalist. It remained now for government to carry on the work that he had commenced and to carry it on with a far larger scope. A thorough scheme of reform devised as much for the good of the public as well

[26] *Morning Post*, 27 April 1831, Notice of Public Meeting on 4 May signed by W B O'Shaughnessy as Honorary Secretary to the London College of Medicine action group.

[27] M. Jeanne Peterson, *The Medical Profession in Mid-Victorian London* (Berkeley, CA, 1978), 8, 26, 124.

as of the profession, in short, a new Medical Act was wanted, and this could only be carried through with the aid of a party in Parliament.[28]

O'Shaughnessy's involvement with the proposed new College was brief but it brought him to the notice of men of influence in London who were in a few months confronted with a new and frightening epidemic with little or no experience to guide them. For O'Shaughnessy, his focus changed to cholera and its pathology, but his long-term future may have been changed by his acquaintance with Joseph Hume through the London College of Medicine campaign. His unstable position in London and Hume's Indian background may have had some bearing on his decision to join the East India Company. It would not have passed him by that Hume returned from his service in India a rich man reputedly with a fortune equivalent today to approximately four million.

Soon after the start of the London College campaign, it was confirmed that cholera had reached England, first appearing in Sunderland. O'Shaughnessy was sent north to investigate the chemical pathology of the disease and his findings would be a major step to understanding for the first time the pathological mechanism of the disease.

[28] S.S. Sprigge, *The Life and Times of Thomas Wakley* (London, 1897), 224.

2

O'Shaughnessy and Cholera,
Intravenous Saline and Latta

Sir Richard Evans, a distinguished historian, when writing about epidemics in general and cholera in particular, said this:

> Society evolved ways of coping with constantly recurring aspects of death and disease, but it was far harder to come to terms with sudden and violent visitations of mass epidemics. Here the novelty of its first impact and the severity of its subsequent visitations stimulated governments, administrators, politicians, caricaturists, doctors, statisticians, and private individuals to pour out a vast mass of writings on the subject which historians in the last two decades have not been slow to exploit.[1]

This vast literature on cholera has analysed the social, political and environmental repercussions of the unstoppable march of cholera across Europe, but historiography has paid little or no attention to the pathophysiological analysis and therapeutic recommendations of a young Irish doctor made in Sunderland and Newcastle during the first cholera epidemic of 1831–1832. It is indeed strange that a society becoming ever more scientifically orientated in the succeeding decades largely ignored a science 'first', perhaps because O'Shaughnessy's work was based on chemical analysis and therefore was totally alien to senior physicians who had been brought up on medical theory still dominated by Galenic ideas and were unable or unwilling to change. His work was largely ignored in the decades following and that neglect has not greatly altered.

There are many reasons for this neglect, not least the rapid disappearance of cholera in Europe until the arrival of the second epidemic of 1848–1849, by which time the work of O'Shaughnessy and Thomas Latta, who had put into

---

[1] Sir Richard Evans, 'Epidemics and Revolutions: Cholera in Nineteenth-Century Europe', *Past & Present*, 120 (1988), 123. Sir Richard Evans is a distinguished British historian of nineteenth- and twentieth-century European history; his seminal work on cholera is *Death in Hamburg, Society and Politics in the Cholera Years* (Oxford, 1987). Recent experience with Covid and social responses to disease and death confirm Sir Richard's opinion.

practice in Edinburgh his recommendations to use intravenous saline had been forgotten, even in the city where first it had been used. It is surprising that this was so for the Edinburgh school of medicine was important in the development of O'Shaughnessy's thinking and especially its chemistry teaching, but it appears that this did not carry much weight when the second epidemic arrived. During his time as an undergraduate, practical chemistry as part of the medical curriculum was being taught mainly in the extramural schools as an adjunct to pure chemical theory and there is no question that O'Shaughnessy benefited from this as his work in Sunderland showed. In his assessment of the development of chemistry, Farrar suggests that:

> chemistry had reached that interesting point in its development where only a small stock of theoretical ideas had to be mastered before embarking on fruitful practical work. The acquisition of technique was more essential, and took longer, than the learning of the necessary basic theory.[2]

In the context of chemical analysis, it is important to look closer to home rather than to Europe, where developments in chemistry were rapid, but to Edinburgh where O'Shaughnessy studied under Professor Thomas Hope (1766–1844), who is generally acknowledged to have been one of Europe's greatest lecturers in chemistry. The theoretical foundation acquired by O'Shaughnessy was therefore firmly in place and was reinforced by lessons in practical chemistry in the extramural school under David Boswell Reid (1805–1863), physician and chemist, and Edward Turner (1796–1837), both experts in their field. It is significant that while still in Edinburgh only one year after qualifying in medicine and acting as an assistant to William Pulteney Alison (1790–1859), Professor of Medicine, he had already published papers on chemical and forensic analysis in *The Lancet*. These early works are examples of his self-confidence and his exceptional ability in practical chemical analysis. Moreover, he had the skills identified by Prout as essential if there was to be any advance in the knowledge of disease pathology and treatment. William Prout (1785–1850), who has been described as 'a consummate analytical chemist', wrote in May 1831 on the science of chemistry:

> and it must be fairly confessed that physiology and pathology have derived much less advantage from this branch of knowledge than might have been expected…. while chemistry was little more than a branch of national philosophy and confined to those who had not studied physiology, what could be expected of it.

---

[2] W.V. Farrar, 'Science and the German University System, 1790–1850', in M. Crosland, (ed.), *The Emergence of Science in Western Europe* (New York, 1976), 185.

His next words were prescient, saying:

> I will venture to predict, that what the knowledge of anatomy at present is to the surgeon, in conducting his operations, so will chemistry be to the physician, in directing him generally, what to do and what to shun; and in short, in enabling him to wield his remedies with a certainty and precision of which, in the present state of his knowledge, he has not the most distant conception.[3]

This prediction was to be confirmed even sooner than he expected when cholera appeared in England and for the first time in history the immediate cause of death was subjected to detailed chemical analysis. Sunderland in the north of England has the distinction that in this seaport a truly scientific approach to the pathophysiology of a disease was first realised by an unknown young physician, Dr William O'Shaughnessy.

This chapter will demonstrate the contrast between the rational approach of O'Shaughnessy, whose chemical analysis was based entirely on scientific principles, and the condemnations voiced by many less enlightened colleagues. It will be argued that his medical education in Edinburgh, which grew out of the empiricism of Enlightenment thinking, encouraged him in logical thought processes although the seeds of enquiry were already planted in a receptive mind by his schooling in Ennis and later in Trinity College Dublin.

The inevitability of cholera appearing in Britain despite having the doubtful security of being a collection of islands forced the government to act. The Privy Council in June 1831 reconvened the Board of Health which had been inactive for decades having been set up first in 1805 in response to the threat of yellow fever. This new consultative Board was replaced in November by the Central Board of Health also based in London and local boards around the country. In August 1831, the Central Board of Health in London rushed out papers on the disease they first called cholera spasmodica, writing on how it had appeared in India and that following ancient trade routes it spread to Russia. Included in this pamphlet was a lengthy section on treatment about which the Board based its knowledge on the experience of surgeons in the East India Company medical service who had seen and treated the disease from the time of its appearance in Jessore, Bengal in 1817. Contributors to this publication included the Privy Council, the Central Board of Health, the Colonial Office together with information from Dr David Barry and Dr William Russell,

---

[3] William H. Brock, 'William Prout 1785–1850', *Oxford Dictionary of National Biography*, (Oxford, 2004); Prout studied medicine in Edinburgh taking his MD in 1811 and was described by Brock as 'a consummate analytical chemist'; William Prout, 'Application of Chemistry to Physiology, Pathology and Practice', *London Medical Gazette*, 28 May 1831, 8, 257–285, 258, 268.

the Board's envoys sent to St Petersburg by the Privy Council in June 1831. The public were told that the Board of Health had already submitted to His Majesty's Privy Council a code of regulations to be adopted in the event of the disease now prevailing in Russia being ascertained to have spread into his Majesty's dominions, and 'herewith transmits a history of it as it appeared in India and in Moscow, together with the modes of treatment adopted in the former country'. The description of the clinical features of cholera taken from reports from medical officers in India and Russia is detailed and horrific.

> The attack of the disease in extreme cases is so sudden, that from a state of apparent good health, or with the feeling only of a trifling ailment, an individual sustains as rapid a loss of bodily power as if he were struck down or placed under the immediate effects of a poison; the countenance assuming a death-like appearance, the skin becoming cold and giving to the hand (as expressed by some observers), the sensation of coldness and moisture which is perceived on touching a frog; by others represented as the coldness of the skin of a person already dead. The skin is deadly cold and often damp, the tongue always moist, often white and loaded, but flabby and chilled like a piece of dead flesh. The voice is nearly gone; the respiration quick, irregular, and imperfectly performed. The patient speaks in a whisper. He struggles for breath, and often lays his hand on his heart to point out the seat of his distress. Sometimes there are rigid spasms of the legs, thighs, and loins. The secretion of urine is totally suspended; vomiting and purgings, which are far from being the most important or dangerous symptoms, and which in a very great number of cases of the disease have not been profuse.[4]

The phrase 'immediate effects of a poison' is apt bearing in mind that O'Shaughnessy approached his forensic analysis of cholera as if he was applying the tools of toxicology to the question, which in a sense is exactly what he was doing. To the present-day enlightened reader, the phrase has meaning as it now known that a toxin, a poison, from a bacterial infection is the immediate cause of the disease. Dread and helplessness must have been the predominant reactions amongst physicians and lay persons alike when reading the words of the Board of Health in 1831. In India, where doctors had been dealing with cholera for a decade, it was inevitable that a sense of powerlessness in the face of this new and awful illness became common among medical officers where 'a feeling of disappointment and almost despair seems to have dispirited the medical officers'. The report summarised the

---

[4] Wellcome Collection, *Papers relative to the disease called cholera spasmodica in India, now prevailing in the north of Europe: with letters, reports and communications received from the continent* (London, August 1831), 2.

treatments favoured by practitioners in India which by their number and oddity indicate the powerlessness of the unfortunate practitioners forced to deal with a disease which their training had not qualified them to understand or treat. They frequently had recourse to bleeding.

The Lancet in an editorial in early November discussed the opinion of Dr A. Brierre de Boismont, a distinguished French physician, who had written:

> The blood in patients affected with cholera undergoes remarkable changes, it becomes black, thickened, viscous and frequently forms a compact mass, separating with great difficulty into serum and coagulum... it is a source of much regret that no <u>satisfactory</u> analysis has yet been made of the blood in cholera. [original emphasis][5]

There are two points to be taken from the long detailed *Lancet* editorial comment: the first is the reference to the lack of a satisfactory analysis of the blood in cholera. Wakley almost certainly was the author of the editorial which quoted de Boismont's words, planting the idea of blood analysis in the editor's mind and perhaps a suitable candidate to carry out such a task. The second point is that on the next page is a review of Bell's *Treatise on the Cholera Asphyxia* praised as a work of sagacity. Bell believed firmly in the importance of venesection 'to relieve the heart and internal organs from a proportion of that deluge of black blood in which they may be said to be drowning'. In the space of a couple of pages is revealed the quandary: the blood is the focus, to be removed at all costs, but with a suspicion lingering that this was not the whole story with more to be discovered.[6]

Even those who were enthusiasts for bleeding admitted that on occasion they were unable to explain why or do explain away their failures. Hamilton Bell wrote:

> Those who are disposed either less favourably towards bleeding, or to condemn it altogether, object, that if the circulation is in a condition to admit of free bleeding, the case is a mild or favourable one, and would probably yield to other remedies. There is no doubt that fatal collapse has sometimes followed even large bleeding... extracts from the reports of those surgeons who have had most experience of the treatment of the disease; which are not only illustrative of the importance of the remedy but prove

---

[5] Editor, 'History of the Rise, Progress, Ravages, etc of the Blue Cholera of India', *The Lancet* (19 November 1831), 255–256; Alexandre de Brierre de Boismont (1797–1881), a French physician, studied the cholera epidemic of 1831 in Poland and concluded that blood analysis was needed.

[6] *The Lancet*, 1 (1831–1832), 255–276, quoting George Hamilton Bell, *Treatise on the Cholera Asphyxia* (Edinburgh, 1832).

the difficulty to which the usual doctrines on the subject of bleeding expose the practitioner, in his endeavours to explain the rationale of bloodletting in Cholera.[7]

There was no rationale for bleeding unless based on a persistent belief in the theory of four humours, a theory which was beginning to be questioned by many doctors even if some adapted the theory to more modern concepts of physiology. From a twenty-first-century perspective the fact that observant medical men persisted with the practice seems astonishing.

Of course, a baffling variety of other remedies were used. It appears that almost all practitioners in the first instance administered opium and as soon as vomiting stopped gave purgatives in which calomel was commonly the principal ingredient, often combined with more opium. A variety of purgatives were employed: croton oil, scammony, rhubarb, jalap, senna, salts, magnesia and castor oil. The conclusion at the end of this advisory section was that 'almost every plan seems to have had its success or its failure', an entirely fair assessment of treatments most of which had no scientific basis whatsoever for its use. However, there was one remedy which was described as the most uniformly successful, 'when it could be used', and that was bleeding 'even in cases when the pulse was scarcely perceptible at the wrist'.[8] Later the Board's official papers allotted a section to treatment in which there were no fewer than three pages devoted entirely to bleeding where the advantages are described at length, highlighting the change in pulse, respiration and the colour of the blood from black to a more normal red. The amount of blood to be removed is recommended: 'eighteen, twenty-four or thirty ounces', the former being close to 500 ml and the latter to almost one litre: in this latter case, it was noted apparently that there was soon observed 'an almost imperceptible pulse rising in power and becoming strong'. These quantities of blood removed will be discussed later when considering O'Shaughnessy's scientifically based recommendations and Latta's remarkable practical demonstration of the use of intravenous saline, treatment that is the complete opposite to that suggested by the Board.[9] But the members of the Board were not alone in their enthusiasm for blood-letting; Indian medical officers were equally enthusiastic and of course the Board was to a great extent reliant on their experience. One such was Dr James Annesley (1780–1847), an Irish-born East India Company medical officer attached to the Madras medical service who wrote on cholera in 1825 claiming that 'bleeding should never be lost

[7] George Hamilton Bell, *Treatise on the Cholera Asphyxia* (Edinburgh, 1832), 141–142.

[8] *Papers relative to the Disease called Cholera Spasmodica*, 8.

[9] *Papers relative to the Disease called Cholera Spasmodica*, 15–17.

sight of'.[10] Alexander Turnbull Christie (1801–1832) thought likewise but his bafflement and admission of ignorance as far as mode of transmission was concerned is in stark contrast to his absolute belief in bleeding writing that 'it is almost universally admitted that blood-letting is one of the most powerful remedies we possess'.[11]

The Board stressed the effectiveness of the quarantine measures they had instituted boasting of 'the security which has been obtained by quarantine' and inexplicably claiming that 'there is no immediate urgency for making public the rigid rules originally laid before the Privy Council by the Board'.[12] Whether this opaque remark concerned quarantine rules is uncertain but the element of secrecy in this document is not untypical of the Board's reluctance to admit that they were powerless to prevent the disease invading Britain; as will be shown, their confidence in quarantine measures was misplaced and was soon to be exposed.

Despite the Board's boasted excellence of the measures now in force the appearance of cholera in Sunderland was inevitable even although quarantine had been extended to include all ships arriving from Baltic ports where cholera was now rife. This misplaced confidence in quarantine measures merely encouraged Sunderland's men of business in their self-deception. The appearance of cholera in Sunderland led to panic and confusion in the town with indiscriminate verbal attacks on the medical profession and on central government; on the one hand, there was pressure on the town's officials to minimise the spread of the epidemic by public health measures and, on the other hand, a reluctance to even accept that cholera was present in the town and spreading.

Local reports from Sunderland confirmed that there was considerable laxity in quarantine control, even extending in certain quarters to actual antagonism towards such measures, antagonism largely based on alarm as to the effect on trade. When stories began to circulate about sailors from a ship out of Hamburg who were reputed to have been seen ashore drinking in local pubs,

---

[10] James Annesley, *Sketches of the most prevalent diseases of India: comprising a treatise of the epidemic cholera of the east; statistical and topographical reports of the disease in the different divisions of the Army under the Madras Presidency* (London, 1825), 167; James Annesley (1774–1847), studied at Trinity College Dublin and spent almost four decades in the Madras medical service, FRS 1840, knighted 1844 and died in Florence in 1847.

[11] Alexander Turnbull Christie, *Observations on the Nature and Treatment of Cholera and on the Pathology of Mucous Membranes* (Edinburgh 1828), 10. It is likely that O'Shaughnessy and Christie knew each other as contemporary matriculated students in Edinburgh.

[12] *Papers relative to the Disease called Cholera Spasmodica*, iv.

the hopelessness of control and the feeble nature of so-called quarantine became clear. A sense of foreboding was expressed succinctly, yet diplomatically, by a writer from Newcastle, who had attended a meeting of the Sunderland health board attended by the leading medical men and magistrates who were reluctant to admit that cholera was present in the town. The writer commented on the presence in the river of a vessel from Hamburg anchored near to the home of the family who had died from the illness and was clearly suspicious that this may have been the source.[13]

The resistance to quarantine appears to have been widespread in and around Sunderland: Dr Brown, who was personal physician to the Marquess of Londonderry, a local landowner with coalmining interests, wrote the following letter, published in several newspapers alongside a letter from the Marquis: 'the shipowners and merchants are in an uproar and are about applying to government to have the restrictions removed. There is a ship of war in the roads to prevent the craft from communicating with the adjacent coast. Vessels from here are subject to fifteen days quarantine.'[14] A London newspaper's correspondent in Newcastle commented on the situation in Sunderland: 'concealment is the order of the day. Dr Daun sent to Sunderland by Government has the greatest difficulty in procuring information.'[15] *The Times* reported from Sunderland of a meeting in the town on 10 November 1831: 'at a numerous meeting today our medical men were severely censured for the unnecessary and destructive alarm they have created...' The belief among the town's officials was that confirming the arrival of cholera in the town was causing harm and even panic. It is remarkable that the demands of trade were of greater consequence that sickness and death and that the meeting described by *The Times* took place sixteen days after the first case was diagnosed! The degree of denial in the town was staggering. Criticism was next directed at the government who were blamed for aiding the spread of the disease: 'our government has not benefited from experience; they have sent some hundreds of ships from places in the Baltic admitted to be infected...'. Those who believed that cholera was contagious supported quarantine measures as proposed by the Board of Health, whereas non-contagionists were against quarantine, thus the debate became more commercial and political than medical. More than a fortnight had passed since the first case had appeared in Sunderland and yet the Central Board of Health remained ignorant of the true situation and the people of Sunderland were kept in the dark.

The Times correspondent went on to quote Dr Christie of the Madras medical establishment, who had analysed the dejections from cholera victims

---

[13] *Hull Advertiser*, 11 November 1831, p. 3, cols 2, 3, 4.
[14] 50 *Nottingham Journal*, 19 November 1831, p. 1, cols 5, 6, 7.
[15] *Weekly Dispatch*, 13 November 1831, p. 2, col. 2.

finding certain chemical characteristics in his work on dejections, partially resembling that of O'Shaughnessy, his erstwhile student contemporary in Edinburgh:

> The most recent experiments on the fluids of cholera that I am aware of, were performed during the last week in Newcastle, by Dr. O' Shaughnessy, of London, who intends, it is said, to deliver a lengthened report of his analysis to the College of Surgeons. A gentleman in Newcastle writes to me, that Dr. O'Shaughnessy has noted a great loss, in the severe cases, of the watery portion of the blood, all the soda it naturally contains, and a large proportion of its other saline ingredients. In the dejections, on the other hand, he detected the alkali in large quantities and also the salts in which the blood was deficient. Dr. Clanny, of Sunderland, has been engaged in a similar investigation; but the results of his research are not, as yet, made public. In order that the preceding, when published, may be easily compared with the results of an Indian analysis, I shall partly quote Dr. Christie of the Madras Medical establishment on the dejections. 1 From a careful examination of the cholera secretion taken from the stomach and intestines of several individuals that died of the disease, I found that it had the following chemical characters and composition. It does not affect litmus or turmeric papers. It becomes of a dark grey colour when mixed with calomel. It consists of two substances, the one a transparent fluid, the other an opaque white coagulum. The former is perfectly soluble in cold water, which enables us easily to separate it from the latter, which is quite insoluble. This separation, which often takes place spontaneously, may be considered the first step towards the analysis of the secretion, in the same way that the coagulation and separation of the crassanmentum form the first step towards the ascertaining the nature of the blood. 1. We must conclude that the cholera secretion is not merely an increased natural secretion of the mucous membranes, but that while it is increased it is also vitiated. The circumstance of its not affecting vegetable colours, proves that there is no free acid in the secretion, and thereby shows that Dr. Ainslie's views of the disease cannot be maintained.[16]

On 26 October, a sixty-year-old keelman named William Sproat, suddenly taken ill, was treated by an army surgeon attached to a local regiment, who pronounced the illness to be cholera, of which he had experience during army service in Mauritius where the disease had spread by sea from India. The following day, the son and granddaughter of Sproat became ill with cholera and were taken to the infirmary, where the daughter recovered but the son died, as did the father. Two further deaths from cholera took place

---

[16] *The Times*, 29 December 1831, p. 3, cols 1, 2.

on 31 October – significantly one of the men dying was also a keelman.[17] On 1 November, a nurse at the infirmary who had helped remove the body of the younger male Sproat died only six hours after she became unwell.[18] The Central Board in London was not informed of Sproat's death until four days after the event when James Butler Kell, the local army surgeon with experience of cholera, went over the head of Dr Clanny of the local Board of Health and notified the Central Board in London who were now forced to admit that cholera had struck in the town. Belatedly, they enforced quarantine on ships sailing from Sunderland, a move that severely affected local trade, angering a group of local businessmen who proceeded to form an 'anti-cholera' party.[19]

The Times correspondent, writing from Sunderland, described graphically the horror of the illness thus:

> During the most torturing attack, and when the cramps have wrung the patient to the expression of apparently feminine [*sic*] weakness, the eye of the sufferer is always dry, and his agony cannot find utterance in tears. His voice, also, being generally impaired in the progress of the symptoms, he speaks only in a whisper, or complains involuntarily in a low melancholy moan. The changes which the blood undergoes in the mortal struggle have been the subject of particular remark and no feature of cholera is found more frequently present than these changes in the circulating fluid…. The blood manifests a strong tendency to become darker in colour and thicker in consistency trickling forth from the veins in drops or in a slow unbroken stream and then coagulating in the receiving vessel into one uniform mass like jelly with scarcely a trace of the serous fluid common to health.[20]

Even *The Times* correspondent emphasised the dark thick blood such was the commonly held belief that this was the cause of the disease. The majority of medical practitioners in Britain believed it, a belief that had the support of the medical hierarchy and was repeated by the Board of Health. It would take a clear-thinking young man unimpressed by power or dogma to look at the problem scientifically: O'Shaughnessy.

---

[17] Keelmen were boatmen on the Rivers Tyne and Wear who used flat-bottomed boats to bring cargo to and from ships downriver, unable because of shallow water to sail further upriver.

[18] *Observations on Cholera made during a Visit to Sunderland undertaken by Direction of the Birmingham Town Infirmary Board of Health in the Months of November and December 1831*, by George Parsons, Surgeon to the Birmingham Town Infirmary (Birmingham, 1832).

[19] *UK Parliament: Cholera in Sunderland*, www.parliament.uk/about/living-heritage/transformingsociety/towncountry/towns/tyne-and-wear-case-study/introduction/cholera-in-sunderland/ (accessed 8 July 2024).

[20] *The Times*, 29 December 1831, p. 3, col. 1.2.

Administrative differences, negative scientific investigations and local antagonism to quarantine apart, the most significant event resulting from the outbreak was the arrival in the town of Dr O'Shaughnessy to carry out chemical analyses of the blood and excreta of cholera victims. It was into this maelstrom of denial, half-truths, and irrationality that O'Shaughnessy prepared to carry out his rational scientific analyses, seemingly unperturbed by the chaos around him and the danger from infection to which he was subjecting himself.

There is evidence that as early as September 1831 O'Shaughnessy had begun to take an interest in the chemistry of the blood, not only carrying out experiments personally but also reviewing the literature, to the extent of writing to the press criticising some of the conclusions of Dr William Stevens (1786–1868), a fellow physician. This interest had been triggered by the cholera epidemic now sweeping through Europe.[21] The discussion and disagreement between O'Shaughnessy and Stevens centred around the difference in colour between venous and arterial blood and what was responsible for this difference, a debate which had arisen because of the thick dark venous blood found in cholera, still a new disease in Europe. The dark tarry venous blood was thought, in accordance with ancient humoral theory, to be one of the main causes of death in cholera and its removal by bleeding was considered to be an essential part of the treatment.[22] Nalin asserts that Stevens made both pathophysiologic and therapeutic errors having no 'quantitative concept of cholera patients salt and water losses or of their quantitative I/V. replacement'. Not only did he make mistakes in his analyses, but Nalin suggests, as did other critics including Sir David Barry and O'Shaughnessy, that some of his so-called cholera patients did not in fact have cholera.[23] Barry and O'Shaughnessy together visited Coldbath Fields Prison to investigate possible cases of cholera, *The Lancet* reporting on O'Shaughnessy 'whose report on the Chemical Pathology of cholera entitles him to such high consideration in everything connected with what has been lately denominated the saline treatment of that disease'.

A belated attempt to enforce quarantine including restrictions on ships sailing from Sunderland proved highly unpopular, severely limiting local trade with the result that local businessmen of the anti-quarantine movement

---

[21] William Stevens, *Dr. Stevens's treatise on the cholera, extracted from his work entitled Observations on the healthy and diseased properties of the blood* (New York, 1832); William Stevens, *Observations on the Blood* (London, 1830); *Observations on the Nature and Treatment of the Asiatic Cholera* (London, 1853), in which he revisits the IV saline debate.

[22] *Evening Mail*, 26 September 1832, p. 3, col. 3.

[23] David R. Nalin, 'The History of Intravenous and Oral Rehydration and Maintenance Therapy of Cholera and Non-Cholera Dehydrating Diarrhoeas: A Deconstruction of Translational Medicine: From Bench to Bedside?', *Tropical Medicine and Infectious Disease*, 7.50 (2022), 1–28, 4.

persuaded doctors to retract their initial diagnosis of cholera.[24] This apparently local 'problem' very soon became a national one as the epidemic spread and it was now that O'Shaughnessy was asked independently, not it seems by the Central Board, to go to Sunderland to investigate the chemical pathology of the infection. O'Shaughnessy later stated that he was sent north at the request of a Vice President of the Royal College of Surgeons in London but no proof that this was the case can be found; it is strange that this body, in view of its quarrel with Wakley and the publicity surrounding the proposed London College of Medicine, would have made such a request unofficially, even if they were impressed by O'Shaughnessy's scientific expertise.[25] In view of his recent criticisms of the College and his position as secretary of the body that existed to oppose the College's closed shop, this request appears unusual but perhaps proves how well regarded were O'Shaughnessy's skills as a forensic pathologist. Perhaps Wakley himself, knowing well his colleague's capability, asked him to go to Sunderland but again there is no evidence. A more plausible explanation is that the London Board of Health encouraged by David Barry and William Russell, the Board's Russian envoys, asked him to go north. Little did anyone imagine just how remarkably this would work out and how significant would be the involvement of *The Lancet*.

His journey to Sunderland, where he arrived on Sunday 11 December 1831, may primarily have been promoted by scientific curiosity, but any emotional detachment was short-lived once he had seen the frightful terminal stages of the disease. He wrote graphically on the last moments of a young woman:

> The colour of her countenance was that of leaden silver blue, ghastly tint; her eyes were sunk deep into the sockets, as though they had been driven an inch behind their natural position; her mouth was squared; her features flattened; her eyelids black; her fingers shrunk, bent and inky in their hue. All pulse was gone at the wrist, and a tenacious sweat moistened her bosom. In short, Sir, that face and form I never can forget, were I to live to beyond the period of man's natural age.[26]

O'Shaughnessy, in an early letter to *The Lancet* from Sunderland, discussed the proposition as to whether the disease was the same as the one 'to which we have been long accustomed' and based his decision, first on the opinions of

---

[24] *UK Parliament: Cholera in Sunderland*, www.parliament.uk/about/living-heritage/transformingsociety/towncountry/towns/tyne-and-wear-case-study/introduction/cholera-in-sunderland/ (accessed 8 July 2024).

[25] W.B. O'Shaughnessy, *Report on the Chemical Pathology of the Malignant Cholera* (London, 1832), iv.

[26] W.B. O'Shaughnessy, 'The Cholera in the North of England', *The Lancet*, 1 (1831), 40–44. This letter was published on 17 December 1831 having been written four days earlier.

medical men long in practice in Sunderland, secondly on men who came to the town from Dublin, Edinburgh and London and finally on his own experience; he was certain that the disease in Sunderland was totally different to the type of cholera long experienced in Britain. It was, he wrote 'a disease entirely different in its ESSENTIAL *symptoms* from that which I have yesterday and today witnessed in this town' [original upper case and italics], saying that the disease of Sunderland and the indigenous cholera of England, were as dissimilar as 'poisoning by arsenic and by hydrocyanic acid', an interesting analogy referencing his forensic training and experience of poisoning. Writing of the 'horrid want of the least approach to a system of medical organisation for the affording of home attendance...' and because of the prejudices of the poor against the cholera hospital, he was sure that treatment was scarcely ever applied until hours after the first seizure. However, on the other hand, cases diagnosed early often become the objects of 'attention and solicitude to all the practitioners, amateur and official, native and exotic...', the patient having to swallow simultaneously the separate medicines ordered by each of the scores of his physicians. He asserted 'that no system can be effectually tried here, and that no practitioner need repair to Sunderland for any further knowledge of the disease than the symptoms of the cases will afford'. Was this outspoken condemnation the catalyst prompting him to observe cholera scientifically and to recommend a rational treatment far removed from the nostrums hitherto in use?[27]

The first intimation of O'Shaughnessy's results and surprising recommendations for treatment came at a meeting of the Westminster Medical Society on 3 December 1831 when he read a paper entitled 'Proposal of a New Method of Treating the Blue Epidemic Cholera by the Injection of Highly Oxygenised Salts into the Venous System', later published in *The Lancet* on 10 December. It has to be remembered that when he addressed this society, he was only twenty-three years of age, two years qualified in medicine and burdened with the humiliation of having been prevented from practising his profession because of restrictive practices enforced by the London Colleges. These considerations do not seem to have hindered him from speaking out, for at an earlier meeting of the Westminster Medical Society in November, when discussing cholera, Dr O'Shaughnessy expressed his opinion that the disease was contagious, the first occasion when he had ventured an opinion on the topic, the discussion having been triggered by reports of cholera in Sunderland. Dr Granville was certain that it was non-contagious, whereas Dr Gregory, whose only conviction was that it had been sent by Providence, did not express an opinion as to its contagiousness.[28]

---

[27] *The Lancet*, 17, 433 (17 December 1831), letter from Sunderland signed W.B.O'S, 401–404.

[28] *Bucks Gazette*, 12 November 1831, p. 4, col. 6.

Later, O'Shaughnessy wrote: 'The habits of practical chemistry which I have occasionally pursued, naturally led me, as well as others, to inquire whether in the remote causes, the pathology or physiology of this disease, any data could be discovered which might lead to the application of chemistry to its cure.'[29] He criticised the use of blood-letting, pointing out the potential harmful effects of the technique when combined with the already debilitating nature of the disease, a simple enough deduction but one seemingly ignored by those who favoured the practice:

> the causes which interfere with the universality of this remedy; and these I apprehend may be found to reside in the debilitating influence, whether transient or permanent, which the detraction of blood sometimes occasions, and which, when added to the debilitating effects of the of the remote cause, becomes sufficient to overwhelm irremediably whatever vital power still clings to the system.

He goes on to quote the opinion of Dr Christison, 'whose distinguished reputation as an accurate chemist, a rational physiologist, and an able practitioner entitles him to our most unlimited confidence'. Christison in this matter was not resistant to change but Warner pointed out that 'Expertise in the scientific bases of medicine could confer a substitute status to the trappings of gentility and access to patronage enjoyed by elite physicians. Perhaps not surprisingly, it was the elite that was most resistant to change and most opposed to foreign notions.'[30] Some of these 'foreign notions' were surely responsible for O'Shaughnessy's interest in chemistry in Edinburgh, inspired by teachers such as Robert Christison who had studied in France and Germany with masters such as Mathieu Orfila (1787–1853) in Paris, considered to be the father of modern toxicology, and Professor Justus von Liebig in Germany. Although O'Shaughnessy had not studied in France, personal study and research was aided by his fluency in French and his ability to read scientific German.[31]

On 31 December 1831, *The Lancet* published what was essentially a preliminary report of O'Shaughnessy's findings in Newcastle and Sunderland, described by him as 'experimental enquiries'. Quite why he felt it necessary to pre-empt the full report in this way by means of a short letter is not clear:

[29] W. B. O'Shaughnessy, 'Proposal of a New Method of Treating the Blue Epidemic Cholera by the Injection of Highly Oxygenised Salts into the Venous System', read before the Westminster Medical Society, Saturday 3 December 1831, *The Lancet* (10 December 1831), 366.

[30] John Harley Warner, 'The Idea of Science in English Medicine: The "Decline of Science" and the Rhetoric of Reform, 1815–1845', in Roger French and Andrew Wear (eds), *British Medicine in an Age of Reform* (London, 1991).

[31] Study in France did not automatically bestow wisdom as the example of James Craufurd Gregory, to be discussed later, demonstrates.

as far as can be ascertained there were no rival researchers ready to publish but there is a hint that he was aware of the possibility of adverse criticism from nameless antagonists, possibly Dr William Stevens (1786–1868), who had written on the blood in cholera also with suggestions as to treatment. The phrase 'experimental enquiries' tells us a great deal about O'Shaughnessy's approach to scientific problems; later in his career this dedication to science and rational conclusions served him well. Here he emphasised the role of practical chemistry in the investigation, the importance of data and the application of chemistry in the search for a cure.

He proposed treatment which contrasted dramatically with contemporary medical orthodoxy largely convinced by the merits of venesection and removal of blood, believing that the dark thick, tarry blood was the cause of the illness rather than the result. His conclusions were based scientifically on his blood analyses and must have surprised and shocked his listeners at the meeting of the Westminster Medical Society, but as far as can be discovered there are no accounts extant describing the reaction of anyone present. The epidemic in Sunderland, he wrote, was such that 'we find ... the mortality in the cases once termed "*malignant*" is still so great that the experience of the past seems almost valueless for present or future protection'. So much for the detailed advice from the Central Board of Health.

In the letter to *The Lancet* of 31 December 1831 headed 'Experiments on the Blood in Cholera', O'Shaughnessy briefly summarised his conclusions, making six points which he clearly grasped were the most important and most critical findings. Although O'Shaughnessy's report on his analyses of the blood and dejections of cholera victims in the north of England was first given to the Westminster Medical Society in early December soon after his return from Sunderland and Newcastle, he did not present his findings to the Central Board until early January.

> The chemical pathology of the Malignant Cholera, as it prevailed in Newcastle-upon-Tyne, was presented to the Central Board of Health on the 7th of January. Its publication has been delayed to the present time, for the purpose of allowing the experiments therein detailed to be more extensively repeated before their results should be admitted as universal facts in the history of this disease.[32]

[32] *Rules and Regulations proposed by the Board of Health for the purpose of preventing the introduction and spreading of the Cholera Morbus. Published by a Committee of the Lords of His Majesty's Privy Council* (London, 1831), 1; appended to this Report were first, a warning to the public written by Sir Gilbert Blane and second, a History of the Epidemic Cholera from 1817 to 1831.

In a prefatory statement addressed to the members of the Central Board of Health, O'Shaughnessy wrote:

> I have the honour to lay before you an account of some chemical inquiries conducted recently at Newcastle-upon-Tyne, into the composition of the blood and alvine dejections, in the malignant cholera now prevailing in that town and its vicinity. In doing so I have considered it advisable, in order to make my statement more generally intelligible, to separate the strict details of chemical manipulation from the observations by which the results of my experiments must be prefixed and associated. I have therefore drawn up the description of the analytic processes employed, in the form of an Appendix to the general Report; to which I now proceed to solicit your attention and which I shall divide into three principal Sections.

The three sections were, first: 'a concise but careful sketch of the exact state of our present knowledge of the chemical composition of the blood in the normal or healthy condition; secondly, an account of all 'the analytic inquiries yet instituted on the chemical pathology of the malignant cholera'; thirdly, 'I will inquire into the extent to which these investigations entitle us to form pathological or therapeutic conclusions'. His report including appendices extended to seventy-two pages of detailed chemical analyses and a considered review of similar investigations carried out by colleagues in the UK, France, Germany and Russia.[33]

Forensic pathology is the discipline which uses scientific methods in the investigation of crimes, most commonly of murder. O'Shaughnessy approached the problem of why it was that people died from cholera exactly as if it was an exercise in forensic pathology, applying scientific methods, in this case chemical analysis, to disease pathology. His careful step-by-step review laying out his conclusions was forensic in its approach. Crime of course was not a factor in cholera, but he looked at the cause of death exactly as if crime was involved; whether he did this consciously or not, it proved an immensely useful approach. O'Shaughnessy as a medical student in Edinburgh attended the classes of Professor Robert Christison (1797–1882), an expert in forensic pathology especially toxicology; Christison studied in Paris observing the methods of Matthieu Orfila (1787–1853), generally accepted as the father of forensic pathology. On his return to Edinburgh, he was appointed at age twenty-five to the Regius Chair of Medical Jurisprudence and Medical Police (public health) in the University of Edinburgh in 1822, introducing many of Orfila's methods, particularly chemical analysis, to his students at the same time promoting forensic toxicology. His Edinburgh successors in toxicology wrote that [he] 'was clear that toxicology was the most promising subject

---

[33] O'Shaughnessy, *Report on the chemical pathology of the malignant cholera*, 2.

for bringing [his] chair, and medical jurisprudence itself, into notice'.[34] He published on analytical and experimental aspects of arsenic, opium and oxalic acid amongst others, but had little experience of poisoned patients, other than in his capacity as a physician in the city's infirmary.[35] His skills were primarily at the service of the law, but his *Treatise on Poisons*, which appeared in 1829, the first of four editions, was widely acclaimed. Orfila understood the critical importance of chemical analysis in forensic examinations and developed sophisticated methods for the detection of poisons in the body.

O'Shaughnessy later reported in Bengal on cases of alleged poisoning all of which had been referred to him within the previous year in his capacity as 'Chemical Examiner to the Secretary to Government in the General Department, dated 2nd April 1841', a position essentially that of a forensic toxicologist. In the extraordinary number of cases in which his chemical and toxicological expertise was crucial, interestingly the first case he reported on and analysed is that of a man who died of cholera, but because a recent altercation with a fellow servant had aroused suspicions of foul play, a post-mortem and a forensic examination of the body was carried out. O'Shaughnessy wrote about this case saying:

> These experiments sufficed to shew, that the changes in the blood and intestinal fluids, now well known to be occasioned by Cholera, had taken place in this case. The serous portion of the blood had passed into the intestinal tract, which caused the alkaline reaction on the test paper; and with the soda of the fluid, the nitric acid formed the crystals of nitrate of soda, obtained in such remarkable abundance.[36]

Whether through modesty or concluding that his cholera analysis during the 1831–1832 outbreak was not relevant in this case, O'Shaughnessy made no mention of his earlier published work at that time.

A London medical journal, in a critical analysis of O'Shaughnessy's seminal report on the chemical pathology, wrote the following:

---

[34] For an account of the long-distinguished career of Sir Robert Christison, see *The Life of Sir Robert Christison, Bart*, edited by his sons, 2 vols, I (Edinburgh, 1885).

[35] A.T. Proudfoot, Alison M. Good and D.N. Bateman, 'Clinical Toxicology in Edinburgh, Two Centuries of Progress', Clinical Toxicology (2013), 1; R. Christison, Treatise on Poisons in relation to Medical Jurisprudence, Physiology, and the Practice of Physic (Edinburgh and London, 1829).

[36] W.B. O'Shaughnessy, *Report on the Investigation of Cases of Real and Supposed Poisoning* (Calcutta, 1841), 1; the importance of expert chemical analysis in these cases was remarked on by a current forensic pathologist, Dr Julie McAdam of the University of Glasgow, having kindly read and commented on O'Shaughnessy's paper: 'What struck me most was the very chemical approach to investigating poisoning at the time of PM (post mortem) and on gross specimens', personal communication, 2023.

40   AN INNOVATIVE PHYSICIAN AND SCIENTIST

It so frequently happened that those who have sought to explain the nature of disease by the aid of the science of chemistry, have bewildered themselves and their readers by fanciful speculations and hasty and unfounded assumptions, that it cannot be subject of astonishment if the practical inquirer looks upon all such investigations with much distrust, and but little hopes of improvement. We deem it proper, therefore, to premise our notice of Dr O'Shaughnessy's very interesting essay, by the assurance that his inquiries have been conducted in the most philosophical and satisfactory manner: he has elicited many most important and novel facts respecting the chemical pathology of the malignant cholera, and in every part, we might almost say in every line of his work, we detect his great anxiety to draw no conclusions but such as clearly and unequivocally arise out of the data he has established.[37]

David Nalin, in an influential history of intravenous and oral rehydration, wrote what should be the final words, first referring to the erroneous conclusions of William Stevens, the incorrect serum analyses of Hermann and the futile attempts of Jaehnichen to lubricate the inspissated blood, saying: 'Then, the more accurate analyses, pathophysiologic and therapeutic paradigms of O'Shaughnessy inspired the daring clinical application of those findings of Latta who advanced rational treatment as far as possible in the absence of sterile solutions and a valid therapeutic method.'[38] Seldom are both men spoken of together in such a way.

Sir Thomas Watson (1792–1882), Bart, FRS, a distinguished London physician, sometime Professor of Forensic Medicine at Kings College and President of the Royal College of Physicians of London between 1862 and 1866, wrote in 1844: 'if the balance could be fairly struck, and the exact truth ascertained I question whether we should find the aggregate mortality from cholera in this country was in any way disturbed by our craft'.[39] Watson wrote these words in a series of lectures published in London, first in 1844 and expanded in several later editions, including one edition from Philadelphia, and at no time did he greatly alter his opinion. His comment may have been

[37] *London Medical and Physical Journal*, New Series, 69 (1832), 389.

[38] Nalin, 'The History of Intravenous and Oral Rehydration and Maintenance Therapy of Cholera and Non-Cholera Dehydrating Diarrhoeas', 1–28, 3.

[39] It is possible that Watson and O'Shaughnessy had been rivals for the chair of forensic medicine at King's College London in 1831, Watson being the successful candidate. The Royal College of Physicians of London archive of Inspiring Physicians, which has replaced Monk's Roll, records that Watson was appointed physician to the Middlesex Hospital in May 1827, and in the arrangements of the University of London, now University College, as a school of medicine, was later nominated to the chair of clinical medicine. He held that appointment for one year only, when his services were transferred to King's College, where he was chosen in 1831 as professor of forensic medicine, holding that office until 1836.

something of an over-simplification but during Watson's life it was true that the medical profession had nothing but palliation to offer in the treatment of cholera. He did refer to the efforts made during the 1832 epidemic to treat cholera on a rational regimen based on the pathophysiological findings of O'Shaughnessy in which fluid replacement was recommended. Sadly, however, O'Shaughnessy's ground-breaking studies, implemented by Latta in Leith and Edinburgh, are not fully credited, perhaps understandably since even later in the century the technique of intravenous fluid replacement was undeveloped and haphazard.[40]

Another historian of medicine, Norman Howard-Jones, in a review of cholera therapy in 1972, wrote the following:

> In the whole of the history of therapeutics before the twentieth century there is no more grotesque chapter than that on the treatment of cholera, which was largely a form of benevolent homicide. To counter persistent vomiting, the physician came to the aid of Nature by administering emetics, and he exacerbated the intractable diarrhoea that was rapidly depleting the sufferer of his vital fluids by drastic purgatives at one end and clysters at the other. The physician also constituted himself the unwitting ally of the Vibrio cholerae by practicing forced exsanguination of patients who were dying from want of circulating fluid. He ransacked the materia medica for new supporting treatments and committed physical assaults upon patients ranging from the boiling water blister to electric shocks.[41]

This indictment of the medical profession was savage but based on the premise that physicians were totally in the dark as to the cause of cholera and in their ignorance clung on to ancient theory which saw blood-letting as the only answer. Fortunately, in Leith and Edinburgh there was a physician, Dr Thomas Aitcheson Latta, who having read O'Shaughnessy's papers in *The Lancet* was determined to put the treatment to the test.

A few months went past before Dr Latta attempted to put O'Shaughnessy's treatment plan into action. In a letter to the Secretary of the Central Board of Health written from Leith on 23 May 1832, Latta explained that his friend Dr Lewins had communicated to him the Board's request for a 'detailed account of my method of treating cholera by saline injection into the veins':

---

[40] Sir Thomas Watson, *Lectures on the Principles and Practice of Physic: delivered at Kings College 1844* (Philadelphia, PA, 1858), 920. In preceding pages, he discusses the use of intravenous solutions but makes no mention of O'Shaughnessy or Latta as the promoters of the treatment.

[41] Norman Howard-Jones, 'Cholera Therapy in the Nineteenth Century', *Journal of the History of Medicine and Allied Sciences*, 27, 4 (October 1972), 373–395, 1.

> Before entering into particulars, I beg leave to premise that the plan which I have put in practice was suggested to me on reading in the *Lancet*, the review of Dr. O'Shaughnessy's report on the chemical pathology of malignant cholera, by which it appears that in that disease there is a very great deficiency both of the water and saline matter of the blood. On which deficiency, the thick, black, cold state of the vital fluid depends, which evidently produces most of the distressing symptoms of that very fearful complaint, and is, doubtless, often the cause of death. In this opinion I am abundantly borne out by the phenomena produced on repletion by venous injection.

The next intimation of this remarkable new treatment came in a letter published in *The Lancet* on 2 June 1832, in effect the same letter written by Latta originally to the Board of Health in London, in which he acknowledges that the treatment was suggested to him in the way that the extract quoted above recounts.[42] O'Shaughnessy's report prompted Latta to treat cholera victims in advanced stages of the disease with intravenous fluid replacement emphasising that patient selection was made according to the experimental nature of the treatment, selecting those who in his judgement were beyond hope of recovery; in other words, his patient selection was determined by the experimental nature of the treatment he was about to use. Although there is no question as to his reliance on the work of O'Shaughnessy, it is quite possible he may also have been influenced by reports that in February 1832, Professor Delpech (1777–1832), physician and surgeon of Montpellier, was visiting Scotland to observe cholera and its treatment, which had not yet arrived in France. Delpech had treated two cholera patients, either in Musselburgh or in Glasgow, by injecting intravenously water containing laudanum, and possibly camphor, but neither patient survived, perhaps not surprisingly, laudanum and camphor being harmful additions to the saline infusion.

First, Latta tried to replace the lost fluid and salts 'by injecting copiously into the larger intestine warm water, holding in solution the requisite salts, and also administered quantities from time to time by mouth...' From these methods he found there to be no permanent benefit and indeed he considered that the unfortunate sufferers' vomiting and purging were aggravated. Latta wrote, 'finding thus, that such, in common with all the ordinary means in use, was either useless or hurtful, I at length resolved to throw the fluid immediately into the circulation'. The injected solution was made up of 'two to three drachms of muriate of soda and two scruples of the subcarbonate of

---

[42] Documents Communicated by the Central Board of Health London relative to the treatment of cholera by the copious injection of aqueous and saline fluids into the veins, *Letter from Dr Latta to the Secretary of the Central Board of Health, London, affording a View of the Rationale and Results of his Practice in the Treatment of Cholera by Aqueous and Saline Injections.*

soda in six pints of water'. He described how 'having no precedent to guide me I proceeded with much caution'. His first patient was an elderly woman who had been given 'all the usual remedies' and had 'apparently reached the last moments of her earthly existence, and now nothing could injure her – indeed so entirely was she reduced that I feared I should be unable to get my apparatus ready ere she expired'. Latta records how:

> having inserted a tube into the basilic vein, cautiously – anxiously, I watched the effect; and injected ounce after ounce of fluid – at first with no visible effect – but then she began to breathe less laboriously and soon the sharpened features, and sunken eye, and fallen jaw, pale and cold, bearing the manifest imprint of death's signet, began to glow with returning animation; the pulse which has long ceased returned to the wrist, at first small and quick, by degrees it became more and more distinct, fuller, slower and firmer.

In the space of thirty minutes after six pints of fluid had been injected, the woman announced in a strong voice that she was now 'free from all uneasiness', her extremities were warm, and Latta, thinking that his patient was now safe, left her in the charge of the hospital surgeon. Sadly, however, the vomiting and purging returned, Latta was not asked to return to the bedside and the unfortunate woman died within five hours.[43] However, he does record graphically the response of a patient on the cusp of death:

> At first there is but little felt by the patient, and symptoms continue unaltered, until the blood, mingled with the injected liquid becomes warm and fluid; the improvement in the pulse and countenance is almost simultaneous; the cadaverous expression gradually gives place to appearances of returning animation, the horrid oppression at the praecordia goes off and the sunken turned up eye, half covered by the palpebrae becomes gradually fuller, till it sparkles with the brilliancy of health, the livid hue disappears, the warmth of the body returns and it regains its colour-words are no more uttered in whispers, the voice first acquires its true cholera tone, and ultimately its wonted energy; and the poor patient, who but a few minutes before was oppressed with sickness, vomiting, and burning thirst, is suddenly relieved from every depressing symptom; blood now drawn exhibits on exposure to air its natural florid hue.[44]

This must surely be one of the most dramatic descriptions in medical history of a patient in extremis being brought back to life. Latta stressed the importance of continuing with the fluid injections, maintaining that 'such remedies must

---

[43] Letter from Dr Latta to the Secretary of the Central Board of Health, London, 23 May 1832, published in *The Lancet* (2 June 1832), 295.

[44] *Letter from Dr Latta to the Secretary of the Central Board of Health*, 6.

44 AN INNOVATIVE PHYSICIAN AND SCIENTIST

be persisted in, and repeated as symptoms demand...', but acknowledged that cure was by no means certain. He considered that failures were caused by giving too little fluid or injecting the fluid at too late a stage in the illness or by the presence of concurrent extensive organic disease, all perfectly valid points.

In an era when chemical knowledge was in its infancy, no attempt was made to standardise the saline solution. Substances were added to the saline often with disastrous results. A Liverpool practitioner injected saline to which had been added egg white, the whole having been filtered through muslin. Initial response to the fluid was excellent, but shortly thereafter every patient experienced the most intense fever with rigors, no doubt a reaction to the foreign protein from the egg white. Finally, with no knowledge of the existence of bacteria and the dangers of introducing infection, several patients who otherwise may have responded to fluid replacement may have succumbed to septicaemia. There are several reports which are very suggestive of such outcomes.

Dr John Mackintosh, physician to the Drummond Street Cholera Hospital in Edinburgh, wrote in 1836 that 'the bold idea of injecting a large quantity of saline solution into the venous system, occurred to the original mind of the late Dr Latta of Leith... in Drummond Street hospital 156 patients were injected of whom twenty-five recovered, a cure rate higher than that for similarly advanced cases'. He stated that 'not one of the patients operated on had a chance of recovery by any other means'. Mackintosh went on to give a comprehensive account of the hospital's experience, describing patient selection, preparation of the solution, method of infusion, results and post-mortem findings, commenting that Latta was 'ably and zealously supported by Dr Lewins'. Patients were given intravenous treatment only after 'every other means had been tried in vain, till the collapse was extreme, and the patient appeared to be in the very jaws of death'.[45]

The hospital cure rate of 16 per cent among those treated by intravenous infusion rises to 19 per cent when the cases reported by Latta are included: sixteen cases treated with eight surviving. This fatality rate of 81 per cent compares with an overall fatality rate of 48 per cent in Scotland which rose to 61 per cent in the age group 40–80; in the 1848–1849 epidemic in Edinburgh the overall mortality was 64 per cent with a very much higher mortality among advanced cases and the aged.[46] Case selection was the reason for the

[45] John Mackintosh, *Principles of Pathology and Practice of Physic*, 4th edn (London,1836), 335–369. The Drummond Street Cholera Hospital was one of three in Edinburgh, the others being Castlehill and Queensferry House. There was also one in Leith, which was a burgh, a port quite separate and independent of Edinburgh at this time.

[46] Neil MacGillivray, 'Food, Poverty and Epidemic Disease Edinburgh 1850–1850' (unpublished PhD thesis, University of Edinburgh, 2004).

high mortality in those treated by saline infusion – Latta emphasised that only patients who had 'reached the last moments of earthly existence' were chosen. Mackintosh was adamant that none of the patients injected with the saline solution had any chance of recovery and referring to similarly advanced cases who had been treated conventionally stated that 'we saw no such miracle out of four hundred and sixty-one cases in the Drummond Street Hospital'. By this he meant that no patient *in extremis* treated by conventional means recovered. Mackintosh was certain that cholera was contagious, a belief that was almost universal and he reflects that he took the post of physician to the Cholera Hospital despite his reluctance to expose his family to the disease; 'during the eleven months that cholera prevailed I made seven visits daily, often ten' and he carried out two hundred post-mortems, each taking at least two hours each. He was making the point that despite such exposure he did not contract cholera and therefore the degree of contagiousness must be slight. It is ironic that Dr Mackintosh sadly succumbed in 1837 to typhus fever thought to have been contracted from a patient.

Both Mackintosh and his colleague, Dr George Meikle, commented on the problem that they had experienced with one patient, a woman who had been given in total 612 ounces of saline (17 litres) and was recovering well until she was removed from the Drummond Street Hospital by the mob and taken to what was described as her 'miserable abode'. Shortly thereafter, Drs Mackintosh, Racey and Meikle himself went there to bring her back to the hospital but once again the mob interfered 'and obliged us to make another hasty retreat downstairs, so that another hour or so elapsed before she was brought to the hospital'. This account is revealing, showing not only the quantity of saline used but also the response of the mob. Cholera riots did occur throughout Europe and there were several in Britain, notably in Bristol and Liverpool, at this time, but if the Edinburgh incident was not a riot, it is an example of how brittle was the relationship between the people and the medical profession. As Cohen points out there were many such riots in Britain, often connected to apprehensions as to body-snatching or 'burking', as the slang term referred to these activities, a word derived from the deeds of Burke and Hare. In Edinburgh, where the two murderers had plied their evil trade, it is surprising that mob violence and rioting was not more prevalent.[47]

The modern physician would find little to criticise in the contemporary defence of the new treatment which cited 'the prevention of stagnation of the

---

[47] George Meikle, 'Trial of Saline Venous Injections in Malignant Cholera at the Drummond Street Hospital, Edinburgh', *The Lancet* (3 August 1832); for a detailed analysis of cholera riots and violence, see Samuel K. Cohen Jr, 'Cholera Revolts: A Class Struggle We May Not Like', *Social History*, 42, 2 (2017), 162–180, and Samuel K. Cohen, Jr, *Epidemics. Hate and Compassion from the Plague of Athens to Aids* (Oxford, 2018).

46 AN INNOVATIVE PHYSICIAN AND SCIENTIST

blood, of the laborious breathing, the burning thirst, the extreme depression of the vital powers, and the chances of aggravating chronic disease, or of producing new organic lesions'. The leading article in the edition of *The Lancet* in which Latta's letter appeared analysed the cases reported from Edinburgh by Latta and other colleagues. The writer concluded that 'the method only failed in one case in which it had been fairly tried – that is, where no organic disease had pre-existed, and where enough of life was left to sanction the least anticipation of success'.[48] In June 1832, O'Shaughnessy wrote to *The Lancet*:

> I beg to state that the results of the practise described by Drs Latta and Lewins exceed my most sanguine anticipations. When we consider that no practitioner would dare to try so novel an experiment except in cases beyond hope of relief by any ordinary mode of treatment and consequently desperate to the last degree, even a solitary instance of recover affords matter for congratulation.[49]

O'Shaughnessy had hoped that the intravenous treatment with saline which he had recommended would be effective and save lives, but Latta's results exceeded even his expectations and it is worthy of note that he refers to 'cases beyond hope'. It is hardly surprising that Latta, who had trained in Edinburgh where evidence-based science was becoming paramount, decided to try this new and radical treatment, which being based on scientific principles, persuaded him that it had merit in contrast to the therapies then in use. At this time in Britain there were no fewer than twenty-four different treatments officially recommended, some eccentric in the extreme and none of any value whatsoever unless in a palliative sense.[50] Many of these so-called treatments were concerned largely with the maintenance of humoral balance, a mode of thought which still dominated medical philosophy. When illness disturbed this imaginary balance, recourse was had to treatment intended to restore the imagined imbalance, namely purging, vomiting, cupping and blood-letting.

Thomas Aitchison Latta MD was born around 1796 in the fishing village of Newhaven, close to Leith, the port of Edinburgh. His father, Alexander, who died in 1807, was a merchant in Leith and an elder in a dissenting Presbyterian congregation, a church whose records have not survived, hence the inability to fix his date of birth and baptism. After his father died, he lived with his

---

[48] Anon., 'Editorial', *The Lancet*, 2 (1831–1832), 284.

[49] *The Lancet* (2 June 1832), letter from W.B. O'Shaughnessy.

[50] E.D.W. Greig, 'The Treatment of Cholera by Intravenous Saline Injections; with particular reference to the Contributions of Dr Thomas Aitchison Latta of Leith (1832)', *Edinburgh Medical Journal*, 53 (1946), 256–263. Included in the twenty-four treatments were such oddities as bastionading the feet and suffocating under a feather bed! Greig was a physician Lieut.-Col. in the Indian medical service.

elder brother, also Alexander, who was a medical student in Edinburgh at the time. When his brother set up in practice in Perth, Thomas moved with him. In 1815, when he himself became a medical student, he returned to Edinburgh. As a young man he was adventurous and while still a medical student he had gone on an expedition in 1818 to Spitsbergen, an island to the north of Norway, as surgeon to Captain William Scoresby, a contemporary Arctic explorer. Evidently, he was considered competent enough to work as a doctor on his own even before graduating in 1819. In the nineteenth century, Leith was a prosperous busy seaport and an attractive prospect for an ambitious young doctor to set up in practice.[51]

It seems certain that before intravenous saline was used in the Drummond Street Cholera Hospital Latta had treated cholera patients on his own initiative, presumably in Leith. Macintosh recalls that 'Dr Latta of Leith saved three patients out of nine in his first set of cases; and in the second set five out of seven. In the Drummond Street Hospital 156 were injected and twenty-five recovered. Not one of the patients operated on had a chance of recovery by any other means.'

According to Dr John Macintosh, his colleague commenced the treatment on 12 May 1832, using Reid's syringe with connecting tubes, great care being taken to ensure that all the connections were air-tight. He records that at one stage twenty-nine cases consecutively died despite all their precautions, this awful record occurring after eight recoveries out of thirty-four cases. He wrote: 'the result alarmed us, and we entered into an anxious investigation to discover the cause of such fatality, and in the end we suspected a faulty state of the tubes; they were cut open, and we had the mortification to discover the spiral wire corroded...' As Venters wrote:

> This was a scientifically reasoned and technically demanding exercise. It required basic surgical competence, practical skill in assembling the relevant apparatus and preparing the infused solution, and rigorous clinical observation during its infusion. Fundamental compassion also guided his decision. He wanted to help his patient and the treatment was a last resort. Time and again his pity and sympathy for his patients shines through in his reports.

In June 1832, the editor of *The Lancet* thanked Dr Thomas Latta for 'the intrepidity, scientific zeal and assiduity he has displayed'. This was in response

---

[51] George Venters, 'Leith in the Time of Cholera: the Story of Thomas Latta', *Hektoen International. A Journal of Medical Humanities*, 7, 1 (2015), https://hekint. org/2017/02/01/leith-in-the-time-of-cholera-the-story-of-thomas-latta (accessed 8 July 2024). Dr Venters, a retired public health consultant, has been of enormous help in encouraging this work on Oshaughnessy, and the discussions we have had over coffee in the National Library of Scotland have been invaluable.

48 AN INNOVATIVE PHYSICIAN AND SCIENTIST

to Latta's letter describing his treatment. He had brought the virtually dead back to life. The recovery of the patient seemed 'more like the workings of a miraculous and supernatural agent than the effect of the interposition of medical science. The case thus alluded to is, we think, one of the most interesting recorded in the annals of our profession.'[52]

Professor Robert Graham (1786–1845), from 1820 the regius professor of botany in the University of Edinburgh, lecturer on clinical medicine and, as the botanic garden was then called, Keeper of the King's Garden, wrote on 28 May 1832 to Dr Nathaniel Wallich, keeper of the Calcutta Botanic Garden, in the following terms:

> We are producing here the most marvellous effects on the worst states of Cholera by injecting marvellous quantities of salt water into the veins. I saw a man today who was far on to the grave yesterday. Into his veins there have been since that time 49 pounds of salt water injected and his eye and feeling and general appearance are now like a man in perfect health.[53] [original underline]

Graham was of course referring to the use of intravenous saline by Dr Thomas Latta in the Drummond Street Cholera Hospital in Edinburgh and was unstinting in his praise for this new therapy. A professor in the university and a fellow of the Royal College of Physicians of Edinburgh, Graham was a prominent physician in the city and as such his support would have been immeasurable – sadly this is the only example of his interest in Latta's treatment, and it lay hidden until recently in the Calcutta Gardens archive. Coincidentally, another aspect of this letter connects both Graham and Wallich with O'Shaughnessy, who attended Graham's botany classes as an undergraduate in Edinburgh and later as an assistant surgeon with the East India Company carried out his first telegraphic experiments in the grounds of Wallich's house in Calcutta and collaborated with him in the preparation of the *Bengal Pharmacopoeia.* It is surprising that there is no written evidence that O'Shaughnessy and Wallich, colleagues in the new Calcutta medical school, ever discussed the cholera treatment innovation of 1831.

Although his first patient succumbed, the change wrought in her convinced Latta of the worth of his treatment and some subsequent cases recovered fully, vindicating his efforts. His local colleagues followed his lead with similar

---

[52] *The Lancet* (2 June 1832), 284–286.

[53] I am grateful to Dr Henry Noltie of the Royal Botanic Garden, Edinburgh for this extract from Graham's letter, the original of which is in the Calcutta Botanic Garden. Dr Noltie is an authority on Scottish medical men and botanists who worked in India and is the author of monographs on men such as Cleghorn and Wight; 49 pounds of water is approximately 22 litres.

success. One, a Dr Lewins, was sufficiently convinced of the major significance of this development as to recommend that Dr Latta tell the Central Board of Health about it and, of course, *The Lancet.*

Nevertheless, it had its detractors from doctors who were still influenced in their thinking by Galen. Latta had the support of local friends and colleagues, particularly Lewins, and was well able to defend himself, but false claims from Dr Thomas Craigie hurt him deeply as is evident from a letter he wrote in *The Lancet* in reply to objections to his treatment. Craigie had chosen to write in a local newspaper that he was the originator of intravenous infusion. Similarly, an anonymous Edinburgh doctor had chosen to misrepresent his results in another paper implying that his success rate was no different from the generality. Professional jealousies may have been the reasons for the disagreements and the fact that Latta was not a Fellow of the Edinburgh College of Physicians must have counted against him – it was not only in London that Colleges had immense power. The repercussions were hinted at in *The Scotsman* newspaper on 7 July 1832 informing its readers that 'we cannot make room for the paper on Dr Latta's saline injections, and we believe there are particular reasons at this moment why controversial papers on the subject should not appear'. The particular reasons might of course have been connected to the disagreement with Craigie but the reference to controversial papers hints at pressure from the authorities, possibly even the Board of Health.[54]

It was John Mackintosh who wrote of his colleague, 'Dr Thomas Latta of Leith, who by his unwearied and unremitting exertions on this occasion, contracted bad health, and died soon after of consumption.'[55] Dr Latta's time in the limelight had drawn to a close. In just over a year after his introduction of his revolutionary treatment, Thomas Latta died on 9 October 1833 from consumption. Given the demands of his clinical work and the burden of disease among the people he cared for, this was no surprise. As Venters has written, medicine complacently turned its back on the door to enlightenment and Latta did not receive the professional recognition that so major an advance deserved and he died without any memorial other than the esteem of his friends and loyal colleagues.

A recent scholar, David Nalin, the doctor who developed oral fluid replacement therapy, in a wide-ranging review of the history of cholera treatment refers to the years before 1831 as the 'apocryphal period when treatment was based on correction of supposed humoral imbalance by bloodletting and purging and other practices sanctioned by longevity and use by ancient hallowed

---

[54] *The Scotsman*, 7 July 1832, p. 1, col. 1.

[55] Macintosh, *Principles of Pathology and Practice of Physic*, 363–364; Macintosh died of typhus in 1837, thus two of the protagonists in the 1832 saline treatment of cholera were dead within five years, both from bacterial disease.

50 AN INNOVATIVE PHYSICIAN AND SCIENTIST

authorities'.[56] The concept of translational medical advances to which Nalin refers is based on the notion of 'bench to bedside', translating the innovation from laboratory to hospital ward. Nalin whose own work was crucial to the development of oral rehydration solutions (ORS) has written that such translational advances rest on three fundamentals: accurate 'understanding of disease pathophysiology', valid therapy for correcting the faulty pathophysiology and a 'clinically effective (and safe) methodology' for treatment delivery.[57] The precise and detailed chemical analysis carried out by O'Shaughnessy in 1831 was complemented by the equally innovative use of intravenous saline by Latta, a treatment which at the time lacked a totally effective and safe method of delivery, a lack which may have delayed by decades the use of intravenous saline until Rogers in Calcutta between 1906 and 1915 scientifically worked out the rationale. The immediate reaction to the publication of Latta's new treatment was generally favourable, although there were exceptions. Leading articles commenting on the use of intravenous saline appeared in the medical press, and there were many letters from practitioners who had also used the intravenous saline treatment: Dr G.F. Girdwood of Islington, writing in August 1832, described seven cases treated by the saline method, five successfully, and exhorted practitioners 'to give this remedy a fair trial'. In his letter Girdwood began by saying that Sir David Barry had 'strongly encouraged him to report his findings to the Central Board of Health' and he continued:

> My own sense of duty towards Dr O'Shaughnessy and Dr Latta would have induced me of my own accord to make public my success, and I only waited till further experience enabled me to make my communication more worth of the authors of this new mode of treatment than it is at present.[58]

The Lancet reported that the blood of one of Girdwood's cases had been analysed by Dr O'Shaughnessy four days after treatment, finding that the quantity of water to be exactly the natural or healthy standard but the quantity of pure salts was less than half the normal standard. Dr J. Wright of the Westminster Cholera Hospital used 'the saline mode of treatment ... but not with sufficient success to induce us to persevere'.[59] Dr James McCabe

---

[56] David R. Nalin, 'The History of Intravenous and Oral Rehydration and Maintenance Therapy of Cholera and Non-Cholera Dehydrating Diarrhoeas', 1–28. Nalin and his colleague Cash in 1968 were instrumental in developing oral rehydration in India thereby rendering intravenous treatment unnecessary in the majority of patients. Nalin in his review paper pays tribute to both Latta and O'Shaughnessy.

[57] Nalin 'The history of Intravenous and Oral Rehydration', 2. *Tropical Medicine and Infectious Disease*, 7, 50, (2002)

[58] G.F. Girdwood, 'Cases of Malignant Cholera treated by Venous Injection', *The Lancet* (11 August 1832), 594.

[59] J. Wright, 'Treatment of the malignant cholera at the cholera hospital, Westminster', *The Lancet* (28 July 1832), 594–596.

of Cheltenham prefaced his critique of the saline treatment writing 'let us now inquire how far the treatment by saline injections into the veins is warranted by, or deducible from, the pathological condition of the blood'. McCabe wrote:

> If we adopt, in our explanation, the humoral pathology, and suppose the cause of these discharges to be the state or condition of the blood itself we must also adopt the language of the older physiologists and refer to an acrimony of the blood and humours. What this acrimony can be unless it be an excess of saline particles in the blood, it is difficult to imagine.

In a few sentences the old humoral theory was used to discredit the use of saline injections and the same sentence reveals the appalling shortcomings of an outdated philosophy when compared with the logical scientific analysis of O'Shaughnessy. This was a moment when ancient medical theory was shown to be totally erroneous. He opines that:

> the loss of the watery and saline parts is the effect of some antecedent cause over which the return of them again into the system by injection can exert no remedial power ... and that an excess and not a deficiency of the watery and saline parts of the blood is a predisposing cause of the disease.

At the same time as the publication of these discussions on saline treatment, letters to the medical press advocated a great many other therapies. Mackintosh listed no fewer than seventy-six different treatments in his *Principles of Pathology and Practice of Physic*, adding that 'it is not even pretended that all the remedies are enumerated'. He pointed out that the list 'would be humiliating to the whole profession were it not remembered how much anxiety and excitement prevailed among medical men at the time that several lost their reason, and many their lives on the occasion'.

Macintosh viewed the saline treatment rationally without prejudice writing:

> I was too old to be led away by any very extraordinary expectations of the results of this practice; and in order that we might err on the safe side, it was determined after deliberate consultation with my kind friend and able colleague Mr Meikle, that no one should be operated upon in this manner till every other means his been tried in vain, till the collapse was extreme, and the patient appeared to be in the very jaws of death.

He soon became convinced of the saline method and succeeded in saving twenty-five out of one hundred and fifty-six injected.[60]

Macintosh's judgement of how desperate doctors were made by the rapidity with which cholera struck and killed, taking desperate measures to save life,

---

[60] Macintosh, *Principles of Pathology and Practice of Physic*, 356, 364–365, 371.

was echoed nearly two hundred years later by Kumar Singh, who wrote succinctly and accurately of the 'attempts by medical men of the age to come up with a successful prophylactic breakthrough in the treatment of cholera', their failures revealing the limitations of contemporary theories. He continued: 'the pharmacopoeia of this heroic age was slim, the barber's blade was sharp and there was no anaesthetic to dull the pain'.[61]

The *London Medical Gazette* printed a letter written on 9 June 1832 by Dr Robert Christison (1797–1882), who had been asked by the Dutch government to advise on the new saline treatment, advice which was to be based on the Edinburgh and Leith experience of treating thirty-seven patients. Christison said that twelve were alive, and of those who succumbed, without exception, they showed at post-mortem signs of extensive organic disease. He was of the opinion that 'the result of these cases is such as to hold out the strongest encouragement to a further trial' and that 'no other remedy has anything like the *immediate* effect of the injection of the saline solution into the veins' [original italics]. He described the effects of the saline injection on a moribund cholera victim, emphasising that these were the *immediate* effects and going on to point out the possible adverse side effects: the risk of air embolism, of phlebitis and the as yet unknown danger of introducing so much saline matter into the blood. Despite these reservations, Christison approved of the principle and was confident that had he been in charge of cholera patients he 'should certainly have given it a trial'.[62]

This letter showed Christison's willingness to give intravenous saline a fair trial, a readiness that does not equate with the criticisms expressed by Dr Robert Lewins of Leith, a friend and colleague of Latta, who claimed that Christison had destroyed any hope of Latta's work being adopted. A private letter from Lewins to Dr William MacLean of the Central Board of Health in London accused several of Edinburgh's leading medical men of being antagonistic to Latta's treatment; he wrote 'the Edinburgh Board of Health, I mean the medical part of them have behaved ill in this matter'. He named Dr Robert Christison and Dr James Craufurd Gregory (1801–1832) as the men responsible. Christison was at this time a leading member of the Edinburgh Board, and one of its two secretaries was Gregory who had recommended that the Board should not use or approve the saline treatment.

Gregory was the third son of James Gregory (1753–1821), who succeeded William Cullen in the Edinburgh chair of the Practice of Medicine. The young Gregory took his Edinburgh MD in 1824 and thereafter studied in Paris for

---

[61] Dhrub Kumar Singh, 'Cholera, Heroic Therapies and the Rise of Homeopathy in Nineteenth Century India', in Deepak Kumar and Raj Sekhar Basu (eds), *Medical Encounters in British India* (Oxford, 2013), 123–124.

[62] *London Medical Gazette*, 10, 7 April 1832 to 29 September 1832, 451, Letter from Professor Robert Christison.

three years, where for a time he was a pupil of Laennec, the inventor of the stethoscope. On returning to Edinburgh, he was soon appointed physician and lecturer at the infirmary and in 1828 was elected a Fellow of the Royal College of Physicians of Edinburgh. Thirty-nine years after Cullen's death he edited in 1829 a new edition of Cullen's *First Lines of the Practice of Physic*:

> with an appendix containing a view of the most important facts which have been ascertained, and which have been adopted, in regard to the nature and treatment of diseases since the death of the author and principles which have been adopted, in regard to the nature and treatment of disease since the death of the author.

Cholera, Gregory explained, was caused by 'an obvious affection of the nervous system... also an uncommonly great and sudden alteration of the circulation and distribution of the blood, [which] is commonly found dark coloured and viscid, probably in consequence of failure of the circulation'.[63] His recommended treatment, largely derived from the experience of doctors treating cholera in India, was to give opium with wine or brandy, calomel and early blood-letting, while admitting that a flow of blood was often difficult to achieve. In all probability, the other man was James Crafurd Gregory, a physician who despite his relative youth appears to have held outdated ideas on the nature of cholera.

Bynum's claim that 'Hippocrates and William Cullen were part of a common medical tradition, separated only by time during which theories of disease causation had hardly changed', emphasises this point. There is no doubt that Gregory was thoroughly traditional in his philosophy and his ideas on human physiology and pathology, ideas based very much on those of William Cullen, a mindset which made his antagonism toward the new therapy inevitable. However, there may have been another dynamic present. At the time of the epidemic, Gregory was seeking to ingratiate himself with other conservative clinicians who would support his candidature for the upcoming chair in medical jurisprudence. In 1832, Gregory applied for the chair of medical jurisprudence, newly vacated by Christison from whom he received a testimonial, dated 23 June 1832. It is unlikely that at such a time the young

---

[63] William Cullen, *First Lines of the Practice of Physic. A New Edition with an Appendix containing a view of the most important facts which have been ascertained, and principles which have been adopted, in regard to the nature and treatment of disease since the death of the author, commenced by the late William Cullen MD and completed by James Craufurd Gregory MD, FRSE* (Edinburgh, 1829); William F. Bynum, 'Cullen and the Study of Fevers in Britain, 1760–1820', *Medical History*, Supplement 1 (1981), 137; Letter from Dr Robert Lewins to Dr W MacLean, Public Record Office, now The National Archives, quoted by Robert J. Morris, *Cholera 1832: The Social Response to an Epidemic* (London, 1972).

Gregory would have been inclined to declare his support for a radical new treatment – even if his beliefs had permitted such a step – beliefs which were firmly rooted in the eighteenth century and, as Bynum has commented, on 'a patho-physiology in which the nervous system was concert master'.

The period during which the saline treatment was in vogue lasted just as long as the cholera epidemic; when the epidemic in Britain ended, there was neither the need nor the opportunity for continued study of the disease or its treatment. The death of Thomas Latta in 1833, and the departure of O'Shaughnessy to India in August of that year, effectively ended research into the treatment of cholera and the chemical pathology of the blood in cholera victims. Of greater importance in the discarding of intravenous infusion was the state of medical knowledge at this time. The prevailing orthodoxy was based on theories of disease where changes in the humours were seen as critical. It was too much to expect physicians brought up on Cullen's theories to view favourably a form of treatment in direct opposition to their beliefs in which bleeding was the mainstay. *The Lancet* in 1831 reviewing 'with admiration' George H. Bell's work on cholera, published in Edinburgh that year, referred to Bell's belief that in cholera there was 'a deficiency of the nervous energy necessary to secretion'. Bell was an enthusiastic advocate of venesection which he believed relieved 'the heart and internal organs from a portion of that deluge of black blood in which they may be said to be drowning'.

During the next cholera epidemic to affect Edinburgh in 1848–1849 it is striking that venesection was used to a greater extent than intravenous infusion. The detailed records of the epidemic maintained in three volumes in the Royal College of Physicians of Edinburgh show that of 739 cholera victims, twenty-seven were given intravenous saline but venesection was used on seventy-eight occasions. The amounts injected were minimal and bore no comparison to those used by Latta sixteen years before and were totally inadequate. It was as if the successful use of saline, limited though it was, had been completely forgotten or dismissed as dangerous.[64] The result of this was that intravenous fluid replacement did not become the standard treatment for cholera or indeed for any condition causing hypovolaemic shock until Rogers in Calcutta used hypertonic saline successfully in the early years of the twentieth century. Latta's pioneering use of intravenous fluid replacement was one example of the physician's craft, rescuing at least some cholera victims from a certain death. There is no doubt that unfavourable reports questioning both the rationale for intravenous saline and its therapeutic value, together with technical difficulties in administering it, contributed to the failure to continue with the treatment and his death in 1833 ensured that he was soon forgotten. If his treatment was

---

[64] Neil MacGillivray, 'Food, Poverty and Epidemic Disease Edinburgh 1850–1850' (unpublished PhD thesis, University of Edinburgh, 2004).

not tried sixteen years later in the city where it had first been introduced, it was unlikely to be tried or viewed favourably elsewhere.

In 1832, Dr Thomas Aitchison Latta of Leith had been the first physician to carry out intravenous injection of saline on a series of patients suffering from cholera. His pioneering work was continued at the Drummond Street Cholera Hospital by Dr John Mackintosh whose words stand as Thomas Latta's epitaph:

> The late Dr Latta of Leith, who by his unwearied and unremitting exertions on this occasion, contracted bad health, and died soon afterwards of consumption. Although Dr Latta's exertions and fate must have been well known to a number of influential men, his grave does not exhibit any monument of public gratitude, nor have his orphan children received any offer of support or protection.[65]

---

[65] Macintosh, *Principles of Pathology*, 371.

3

Bengal Dispensatory and *Cannabis Indica*

This chapter will focus on O'Shaughnessy's innovative pharmacological research in which he studied the properties of several medicaments which had long been used in India but were quite unknown in the West. His research arose from his membership of a government committee selected in 1838 to compile a *Bengal Dispensatory and Pharmacopoeia.* On being appointed as the editor, he explained in his introduction written in October 1841 the purpose behind the enterprise:

> Four years have now elapsed since the appointment of a Committee to examine and report upon the state of the Honourable Company's Dispensary and the possibility of substituting indigenous remedies for some of which are only procurable from other countries at prices which place their use beyond the means of the mass of the population.[1]

The previous explanation for this venture from the Company was that there was a pressing need to save money on importing drugs rather than a beneficent wish to help the native population – or was this the editor's version? Whatever the reason for compiling a dispensatory, it enabled O'Shaughnessy to examine drugs used locally in Bengal and soon led him to *Cannabis indica.* He explained how in his work on this project 'he limited his attention to points of practical utility', avoiding as he put it 'all that merely bore on the literature or controversies of the subject', admitting that he was often tempted to wander into these 'more pleasing fields of inquiry'. Tempting as these were, he kept to his brief examining indigenous medicaments 'of reputation for their medicinal virtues' by both chemical and clinical experiment. 'The febrifuge powers of Narcotine, the purgative Kaladana, the emetic Crinum, the extraordinary stimulant and narcotic Gunja with many other substances were thus successively examined.'[2] The narcotic Gunja or *Cannabis indica* became the focus of his interest, in time demonstrating its therapeutic potential by carrying out

---

[1]   W.B. O'Shaughnessy, 'Introduction', in *Bengal Dispensatory* (Calcutta, 1842), iii. He wrote the introduction in October 1841.

[2]   O'Shaughnessy, 'Introduction', xvi, xviii.

58   AN INNOVATIVE PHYSICIAN AND SCIENTIST

meticulous drug trials including animal testing, a highly unusual method at
the time. His results were published in medical journals, first in India and
later in Britain, arousing considerable interest amongst the medical profession,
new additions to the drug armamentarium being few in the early years of the
nineteenth century. It is remarkable how his cannabis research has continued
to interest clinicians to the present day, an aspect that will be touched on later
and of course the role of cannabis as a medication has yet to be resolved in
many countries worldwide.

> The modern history of medical Cannabis begins in the xix century when
> the scientific method was applied for the first times in the study of its
> pharmacological and toxicological properties. The first, to our knowledge, to
> apply the experimental method in studying Cannabis was the Irish physician
> William Brooke O'Shaughnessy who worked in India, where he conducted
> his experimentations.[3]

The authors of the paper from which the words above are taken wrote a
detailed history of cannabis, tracing its use from ancient times to the present
and in doing so properly identified O'Shaughnessy as the first physician to
employ modern techniques in evaluating its properties. As was apparent from
his careful analytical work in 1831 on the pathophysiology of cholera described
in an earlier chapter, forensic chemical analysis was not his only strength. He
went further with a bold and innovative recommendation for the treatment of
cholera. So it was with his research into cannabis where 'scientific method'
was applied possibly for the first time ever in drug evaluation, but he went
on to make suggestions after trials as to which medical conditions cannabis
could be of potential use. This was not his first foray into drug research; he had
barely arrived in India as an assistant surgeon in the Bengal medical service of
the East India Company, hardly a position of influence or importance, before
he was making a powerful impression with his scientific ideas, experimenting
with plants which might be pharmacologically useful and reporting his findings
in lectures to the Asiatic Society or in that Society's Proceedings.

Many scholars, past and contemporary, believe that in the Arabic and
Persian worlds, physicians in antiquity and in later centuries knew about the
therapeutic properties of cannabis but, as O'Shaughnessy makes plain, he was
unable to discover exactly how it was used, in what dosage and whether there
were unpleasant and unwanted side effects, all questions that are relevant to
drug introduction in the present day; accordingly, he set out to remedy this gap
in knowledge by a series of careful scientific experiments. In his introduction

---

[3]   S. Pisani and M. Bifulco, 'Medical Cannabis: A Plurimillennial History of an
Evergreen', *Journal of Cellular Physiology*, 234, 6 (2019), 8346.

to the *Dispensatory*, he acknowledged the contribution of his indigenous medical colleagues gaining some notion of its potential from discussions with them, for he wrote 'in the popular medicine of these nations, we find it extensively employed for a multitude of affections. But in Western Europe its use either as a stimulant or as a remedy is equally unknown.'[4] An attitude rather different from that espoused by the Anglicists some who saw nothing useful in Indian medicine or *materia medica*; now pragmatism prevailed.

The influence of Ayurvedic and Unani medicine on the British medical practitioner in India has been explored by several scholars, including Professor Dominik Wusjastyk and Professor Poonam Bala, who both conclude that the effects of the anglicising of language and culture on Indian doctors was considerable. Wujastyk, examining the effect of the colonisation by the British, contends that:

> Under first the Moghul and then the British colonial powers, indigenous Indian medicine survived as it always had, mainly through support from patients and the community, but with occasional patronage from the state, with education and practice being devolved and decentralized, often taking place at the family level.[5]

Bala in her doctoral thesis takes a similar position but points out how with Anglicisation and increasing control from the imperial power traditional Western medicine became dominant:

> Initially, it seemed possible that the two systems of medicine could live in peaceful co-existence and mutual accommodation, but gradually, with professional pressure from Britain and State sanctioning in India, Western medicine moved to a dominant position in State provision of medical services. By the end of the nineteenth century, advances in Western medicine undermined the similarities of theory and practice which, earlier, made extensive co-operation seem at least a possibility.[6]

Until the late 1820s, as Anglicisation became the creed, a live-and-let-live attitude existed encouraged by the colonial administration and this applied equally to Indian medicine, evidenced by the establishment of the native medical institutions. Bala in her research is of the opinion that 'the initial

---

[4] W.B. O'Shaughnessy, 'New Remedy for Tetanus and other Convulsive Disorders', *The Lancet* (July 1840), 539.

[5] Dominik Wusjastyk, 'Medicine, India', in Maryanne Cline Horowitz (editor in chief), *New Dictionary of the History of Ideas* (Farmington Hills, MI, 2005), 1411.

[6] Poonam Bala, 'State and Indigenous Medicine in Nineteenth- and Twentieth-Century Bengal: 1800–1947 (unpublished PhD thesis, University of Edinburgh, 1987), v.

60 AN INNOVATIVE PHYSICIAN AND SCIENTIST

policies favoured utilization of indigenous drugs and encouraged training and employment of indigenous medical practitioners. This was facilitated by the similar basis of diagnosis and treatment in Indian and Western medical sciences.[7]

O'Shaughnessy's appointment to the East India Company and arrival in Bengal coincided with this critical period for education in India, when the struggle for educational and language supremacy between the Orientalists and the Anglicists was about to be resolved in favour of the latter. Critical to this change was the arrival in Bengal of Charles E. Trevelyan (1807–1886), later Sir Charles Trevelyan, Bart and Thomas Babington Macaulay (1800–1859), later Baron Macaulay, who altered the tone of the debate aided by the support of Lord William Bentinck (1774–1839), governor of Bengal between 1828 and 1835 when he became the first governor-general of India. It has been suggested that Macaulay's 1835 *Minute on Indian Education* was less important to the question of language than has generally been assumed, perhaps exaggerated both at the time and later, but it seems to have effectively ended the debate in favour of English as the medium of education in Indian secondary schools and in colleges.[8] The leader of the Orientalist side of the debate was Horace Hayman Wilson (1786–1860), who came to Bengal in 1808 as an assistant surgeon with the East India Company but was soon appointed deputy assay master in the Calcutta mint and thereafter began to study Sanskrit and to promote the Orientalist cause. It has been said with good reason that 'in his various capacities, official and unofficial, Wilson kept the government colleges of Bengal as securely Orientalist as he could until his departure for England in 1833'.[9] Wilson later became a close friend and confidant of O'Shaughnessy despite a considerable difference in ages. It could be argued that the relationship was based on a mutual interest in native culture although there is no evidence that O'Shaughnessy was ever outspoken in support for Orientalism.

[7] Bala, 'State and Indigenous Medicine', 8.

[8] Lord William Bentinck's order of March 1835: His Lordship-in-Council is of the opinion that the great object of the British Government ought to be the promotion of European literature and science among the natives of India, and that all funds appropriated for the purpose of education would be best employed on English education alone. His Lordship-in-Council directs that all the funds which these reforms will leave at the disposal of the Committee be henceforth employed in imparting to the Native population knowledge of English literature and science through the medium of the English language.

[9] Natalie R. Sirkin and G. Sirkin, 'The Battle of Indian Education. Macaulay's Opening Salvo Newly Discovered', *Victorian Studies*, 14, 40 (June 1971), 111; Hayman became a friend of O'Shaughnessy as is shown in letters exchanged between them at the time of O'Shaughnessy's election to the Royal Society. Hayman was appointed Professor of Sanskrit in Oxford University in 1833 and FRS in 1834.

The Native Medical Institution was founded in Calcutta in 1822 to teach medicine in the vernacular to Indians, where anatomy, medicine and surgery texts were translated into local languages. The first superintendent was John Tytler (1790–1837), a Scots-born surgeon in the Bengal Service who was a brilliant linguist, fluent in Arabic, Sanskrit, Hindi, Persian and Bengali; unsurprisingly, he was a fervent Orientalist.[10] Islamic medicine was not ignored for classes in the Moslem medical tradition were held from 1826 in the Calcutta madrasa.[11] It is interesting to note that in 1828, Robert Montgomery Martin (1801–1868), an Irish doctor, author and colonial administrator, put forward a plan for a new medical college before the viceroy, Lord William Bentinck, but the idea was rejected at the time 'by the Supreme Government, lest Hindoo sensitivities be injured'.[12]

The move towards Anglicisation with the arrival of Macaulay and Trevelyan, frequently driven by evangelical inspiration, changed this accommodation but O'Shaughnessy as a new arrival shortly to be appointed a professor in the Calcutta Medical College was to an extent shielded by being involved in teaching students who were Indian, some of whom had studied previously in the Native Medical Institution which was abolished only a few years after it had been opened. It is to his credit that he did not ignore indigenous medicine when he began his first investigation into cannabis, first researching its use by native practitioners, nor did he feel other than proud of his students, writing in his introduction to the *Dispensatory*, 'I offer the *Dispensatory* and *Pharmacopoeia* to the students of the Medical College, the only class for whom I presume to consider them a fitting guide.'

In the introduction to his monograph on the properties of cannabis, O'Shaughnessy first lists the many countries where the narcotic effect of hemp was well known, such as the South of Africa, Egypt, Asia Minor, South America, India, Turkey, Burma and Siam. 'In the popular medicine of these nations we find it extensively employed for a multitude of affections

[10] Michael J. Whitfield, 'Dr John Tytler (1787–1837), Superintendent of the Native Medical Institution, Calcutta', *Journal of Medical Biography*, 4 (2021), 184–189.

[11] Anshu and A. Supe, 'Evolution of Medical Education in India: the Impact of Colonialism', *Journal of Postgraduate Medicine*, 62, 4 (2016), 255–259.

[12] Jayanta Bhattacharya, 'From the Inception to the First Dissection Calcutta Medical College 1836', *Doctors' Dialogue* (2023), 2. Presumably the students attending the new college proposed by Martin were to be taught in English. Zhaleh Khaleeli, 'Harmony or Hegemony? the Rise and Fall of the Native Medical Institution, Calcutta, 1822–35', *South East Asia Research*, 21, 1 (2001), 77–104, at 77. David Arnold, *Colonizing the Body: State Medicine and Epidemic Disease in Nineteenth Century India* (Berkeley, CA, 1993), 5. Revd Alexander Duff (1806–1878), Church of Scotland missionary and ardent supporter of the Anglicist project was an important figure in the Anglicist faction.

especially those in which spasm or neuralgic pain are prominent symptoms.'[13] It is something of a paradox that this interest in ancient Indian cures took place at a time when there was a major change in the attitude of the British not only to native Indian languages, but also to native medicine as had been practised for centuries in its Ayurvedic and Unani forms. The abolition of the native schools of medicine less than a decade after their foundation was undoubtedly a blow to many indigenous and East India Company employees who believed that allowing the different languages and cultures to exist side by side was beneficial to all.

An empirical approach to evaluating drug treatment and response was unheard of in early nineteenth-century Britain or India and that he was doing this type of work remote from European academic centres makes his cannabis research more remarkable. It is apparent that having developed good relations with his Indian physician colleagues and students in the new Calcutta Medical College, he became interested in the ancient texts of Hindoo medicine: the Ayurveda. These have been described by Whitelaw Ainslie, an East India Company surgeon, as medical writing of the highest antiquity dating back to the second millennium BCE, and in those ancient texts was documented the use of bhang, derived from *Cannabis indica*, as 'one of five sacred herbs'.[14] In Indian literature, the earliest mention of the word 'bhang' is in the Athar Veda, thought to have been written between 2000 and 1400 BCE. The first mention of bhang as a medicinal plant is in the works of Susrata dating back to the sixth or seventh century.[15] By the tenth century, its narcotic and analgesic properties had begun to be recognised, as were its intoxicating and pleasure-giving effects.

Ancient though the Ayurveda might be, there is even earlier documented evidence of cannabis use in China, confirmed by carbon dating and dating back to 4000 years BCE. However, it was the historical evidence from Hindoo sources that encouraged O'Shaughnessy to look further into the plant's possible uses, undoubtedly taking as his guide the notion that 'the existence of the historic use of a botanical in medicine provides a presumption of sufficient safety and efficacy to justify the investigation of the botanical drug'.[16] If it had been used medicinally for hundreds of years it was unlikely

---

[13] W.B. O'Shaughnessy, *On the Preparations of the Indian Hemp or Gunjah (Cannabis Indica)* (London, n.d.). [Reprinted from the *Transactions of the Medical Society of Calcutta* (1838) and from the *Provincial Medical Journal* (1843)], 3.

[14] Whitelaw Ainslie, later Sir Whitelaw Ainslie, *Materia Medica of Hindoostan and Artisans and Agricultural Nomenclature* (Madras, 1813). In 1826, he published a further enlarged version in two volumes.

[15] The Compendium of Suśruta is a treatise on classical Indian medicine written in Sanskrit about two thousand years ago. It is one of the foundation texts of Ayurveda.

[16] Ethan Russo, 'History of Cannabis as a Medicine', in Geoffrey Guy, Brian

that there were dangerous side effects. Not only did he become intrigued by the ancient history of cannabis, an interest that membership of the Asiatic Society gave considerable stimulus to research with exchange of ideas with like-minded colleagues.

The Society may have been remote from British centres of learning, but it was expansive in its ambitions. In the first issue of March 1832, the *Journal*, a record of the transactions of the Asiatic Society, stated as its mission 'to give publicity to such oriental matters as the antiquarian, the linguist, the traveller and the naturalist may glean, in the ample field open to their industry in this part of the world, i.e. Asia, and as far as means would permit, to the progress of the various sciences at home, especially such as are connected in any way with Asia'. The first editor and Society secretary, James Prinsep (1799–1840), devoted a considerable amount of his time and energy to the decipherment of Brahmi, an activity which became the focus of great scholarly attention in the early nineteenth century, in particular by the Asiatic Society; Prinsep eventually deciphered Brahmi, publishing his results in a series of scholarly articles in the Society's journal in the 1830s.[17] A mission statement suggested that the ambitions of the Society were more focused on literary and linguistic matters than on scientific ones, but the existence of the Medical and Physical Society as part of the Asiatic Society meant that as time went on more medical and scientific papers were being heard and published.

O'Shaughnessy began his research into the properties of cannabis by studying the past literature, especially reading work on native medicines, typically the botanical studies carried out by East India Company surgeons: Whitelaw Ainslie, John Fleming and George Playfair. All three were Scottish surgeons, the latter two having studied botany and medicine in Edinburgh. Fleming wrote 'A catalogue of Indian medicinal plants and drugs, with their names in the Hindustani and Sanskrit languages', which appeared first in *Asiatick Researches* in 1810, a volume that was reissued the same year as a separate publication with emendations and additions.[18] He worked briefly as the curator of the Calcutta Botanic Gardens, acting as a locum until William

---

Whittle and Philip Robson (eds), *The Medicinal Uses of Cannabis and Cannabinoids* (London and Chicago, IL, 2004), 1–16, v; Russo quotes H.-L. Li, 'An Archaeological and Historical Account of Cannabis in China', *Economic Botany*, 28 (1974), 437–448.

[17] Prinsep was a member of the Bengal Dispensatory Committee and was assay master of the mint.

[18] *Asiatic Researches* was started in 1788 by Sir William Jones who had founded the Asiatic Society in 1784. At the time of its foundation, the Society was named the 'Asiatick Society'. In 1825, the society was renamed as 'The Asiatic Society', before in 1832 the name was changed to 'The Asiatic Society of Bengal'. Again in 1936 it was renamed as 'The Royal Asiatic Society of Bengal'. *Asiatic Researches* eventually became the *Journal of the Asiatic Society of Bengal* in 1832.

Roxburgh had recovered from illness and was fit enough to take over in 1807. Roxburgh (1751–1815) was another Scot, born in Ayrshire, who matriculated at the University of Edinburgh in 1771 and studied botany in Edinburgh under Professor John Hope. Roxburgh has been described, with good reason, as the founding father of Indian botany and the greatest botanist since Linnaeus.[19] Through Hope's influence, he obtained in May 1766, although not yet qualified, an appointment as surgeon's mate on one of the East India Company's ships. After making several voyages to India and completing his medical studies at Edinburgh, he was appointed as assistant surgeon in the East India Company's Madras establishment. Later, an association with Sir Joseph Banks led to the publication of *Plants of the Coast of Coromandel*, which appeared in several folio parts between May 1795 and February 1820. In the context of botanicals useful in medicine, it is noticeable that each of its 300 engravings had information on native uses in agriculture, food, and medicine.[20]

Whitelaw Ainslie became a surgeon in the East India Company service in 1788 and devoted the latter part of his life to medical history, specifically that of Indian *materia medica*. O'Shaughnessy reckoned that his *materia* was the best work yet published on this topic with a catalogue of several hundred plants identified and named, 'but scarcely one has been subjected to analysis and very few have been made the object of clinical investigation'.[21]

John Fleming (1747–1829) also studied in Edinburgh, taking his MD there before joining the East India Company as an assistant surgeon in 1768, rising to become President of the Company medical board in 1800. He attended John Hope's botany classes in 1765 and 1766 and knew all the great surgeon naturalists.[22] Fleming could himself be described as a surgeon naturalist having written the first published survey of Indian drugs intended for the use of European doctors newly arrived in the country, in which he discussed botanicals useful in medicine:

[19] Desmond, Ray, 'Roxburgh, William (1751–1815)', *Oxford Dictionary of National Biography* (2004).

[20] *Plants of the Coast of Coromandel* (published under the direction of Sir Joseph Banks), by William Roxburgh, Joseph Banks and Patrick Russell, 3 vols (London, 1795–1820). All three volumes have the most exquisite botanical drawings; for more on the life and work of Roxburgh, see T.F. Robinson, 'William Roxburgh 1715–1815, The Founding Father of Indian Botany' (unpublished doctoral thesis, University of Edinburgh, 2003), later published as Tim Robinson, *William Roxburgh. The Founding Father of Indian Botany* (Chichester, 2008).

[21] B.D. Jackson, revised by James Mills, 'Sir Whitelaw Ainslie (1787–1837)', *Oxford Dictionary of National Biography* (2004); O'Shaughnessy, 'Introduction', xiv.

[22] Henry Noltie, 'A Tangled Calcutta-Caledonian Web: James Kerr, John Fleming and John Hope's Engravings of Asafoetida', *Botanic Stories*, Royal Botanic Garden, Edinburgh (2023), https://stories.rbge.org.uk/archives/37241 (accessed 8 July 2024).

The following catalogue is intended chiefly for the use of gentlemen of the medical profession on their first arrival in India to whom it must be desirable to know what articles of the Materia Medica this country affords. The systematic names of the plants are taken from Willdenow's edition of the Species Plantarum L. with the exception of some new species not included in that work, which have been arranged in the system and described by Dr Roxburgh; who with his usual liberality, permitted me to transcribe their specific characters and trivial names from his manuscript.[23]

Another important text available to O'Shaughnessy would have been *The Taleef Shereef or Indian Materia Medica*, a work translated from the original by George Playfair (1782–1846), MD University of St Andrews, where his father was the principal. Playfair, another Scot, joined the Bengal medical service eventually becoming superintending surgeon of the service and latterly, inspector-general of Bengal hospitals. O'Shaughnessy and Playfair certainly knew each other in Bengal for it is on record that when Lyon Playfair, George's son, came to Calcutta in the late 1830s, he so impressed O'Shaughnessy with his ability in chemistry that he persuaded him to return to Scotland to continue his chemistry studies. After some years studying in Glasgow, Edinburgh and London, the young Playfair went on to study with the great Justus von Liebig in Giessen, where he took his PhD in 1841 and later translated Liebig's works into English at Liebig's request. Lyon Playfair (1818–1898), after a distinguished career in science, academia and politics, became 1st Baron Playfair. The translator of the *The Taleef Shereef*, presumably George Playfair himself, wrote this introduction:

In the course of a practice of upwards of twenty-six years in India, I have often had occasion to regret, that I had no publication to guide me, in my wish to become acquainted with the properties of native medicines, which I had frequently seen, in the hands of the Physicians of Hindoostan, productive of the most beneficial effects in many diseases, for the cure of which our Pharmacopeia supplied no adequate remedy; and the few which I had an opportunity of becoming acquainted with, so far exceeded my expectations, that I determined to make a Translation of the present work, for my own gratification and future guidance.

The Taleef Shereef is essentially a pharmacopoeia of Indian drugs and their uses, published by the Medical and Physical Society of Calcutta in 1833, the society which O'Shaughnessy joined the following year. The work of Ainslie, Fleming and Playfair recording native plants and their therapeutic uses would certainly have been of great interest to O'Shaughnessy and must have had an

---

[23] Ray Desmond, 'John Fleming (1747–1829)', *Oxford Dictionary of National Biography* (2004).

66 AN INNOVATIVE PHYSICIAN AND SCIENTIST

influence on his research. Although these works were botanical in form and intention, as Playfair made clear, due emphasis was given to the medicinal uses of the plants and herbs depicted. These books would have been familiar to O'Shaughnessy, who also had a background in botany having studied the subject in Edinburgh as an essential part of his medical degree. It is a great irony that his connections with Edinburgh medicine, botany and chemistry were proving to be of much greater value professionally in Calcutta than they ever were in London.[24] These volumes would have been an essential source of knowledge in his own survey of native plants and medicines. That the era of colonists interested in indigenous languages was about to be overtaken by one where English was to predominate officially is another irony.

As has been pointed out, the early decades of the nineteenth century witnessed a change in British attitudes not only to Indian languages but also to native Indian medicine as practised in its Ayurvedic and Unani traditions. The abolition of the native schools less than a decade after their foundation was undoubtedly a blow to those who believed that allowing the different languages and cultures to exist side by side was beneficial. O'Shaughnessy appears to have surmounted many of the difficulties posed by these changes and, as Russo rightly points out, 'it was not until O'Shaughnessy, a physician in India who carried out work between 1838 and 1840 that Indian hemp truly came into its own in Western medicine'.[25]

O'Shaughnessy's method of carrying out research had its origins in his studies in the Edinburgh medical school where scientific method and empiricism drew its core value from the disciplines of Scottish Enlightenment thinking. Here, as has been shown in earlier chapters, he had studied under men such as Thomas Hope in chemistry, Robert Christison in forensic medicine and Robert Graham in botany, men whose teaching had taught him to seek logical treatment of diseases and to examine the use of drugs rationally exactly as he had done with his recommendation of intravenous saline in cholera. His undergraduate botanical studies were critical to his developing interest in pharmacology and his approach to drug evaluation was no different, even if based to an extent on chemical analysis.

Whether influenced by Anglicist propaganda or not, many East India Company surgeons had come to see Indian medicine as 'stagnant and inefficacious' possibly because it was interwoven with 'social and religious practices' which in time they had come to look down upon, despising traditional medicine and those who practised it. Increasing evangelical proselytisation by many who held important roles in colonial rule exacerbated this contempt, perhaps

---

[24] George Playfair, *The Taleef Shareef or India Materia Medica* (Calcutta, 1833); Graeme, J.N. Gooday, 'Playfair, Lyon, first Baron Playfair, 1818–1898', *Oxford Dictionary of National Biography* (2004).

[25] Russo, 'History of Cannabis as a Medicine', 1–16.

deliberately. At one time, it had been British policy to encourage the use of indigenous drugs but as pressure eroded and eventually ended the role of indigenous languages so the attitude to medicine changed, the result being that Western medicine became the dominant force.[26]

Thus, in nineteenth-century Calcutta the colonial imposition of the English language and of Western medicine were happening simultaneously. Both these shifts brought about a greater dependence on Britain for higher education and progress, medically and intellectually.[27] Khaleeli quotes Arnold, who suggests:

> that what lay at the heart of colonial interest in native medicine was the discovery and incorporation of new drugs into the Western pharmacopoeia as well as a sinister mission to capture the essence of Indian civilisation and then demonstrate Western superiority.[28]

The rationale as to why the East India Company was eager to find new drugs was partly economic and partly benevolent as was argued by O'Shaughnessy, but there is no doubt that the pressure was on for its surgeons to look for new drugs locally. It had become a matter of economic necessity for the Company to source medication in India where the production of drugs would be cheaper and no longer necessitate the long sea voyage round the Cape. But it is much less certain that there was a sinister mission as suggested by Arnold at the heart of this interest. Nevertheless, the imposition of English as the language of education might well be construed as a mission to disseminate the idea of European superiority, although there were other issues which were important in promoting the change: with increasing proselytisation the compulsion to evangelise and convert the native population to Christianity had become a major part of the mission and the English language was the means whereby this was to be achieved. The attitude of Revd Alexander Duff (1806–1878), a Scottish missionary, may not have been untypical, when addressing the General Assembly of the Church of Scotland, he said: 'the elaborate systems of Hindu learning – geography, astronomy, metaphysics, medicine, law etc – abounding as they do with the grossest imaginable errors and all of them found in the shasters i.e. the sacred books of canonical authority'.[29] Whether

---

[26] Poonam Bala, *Imperialism in Bengal* (New Delhi, 1991), 17, 20.

[27] Zhaleh Khaleeli, 'Harmony or Hegemony? the Rise and Fall of the Native Medical Institution, Calcutta, 1822–35', *South East Asia Research*, 21, 1, (2001), 77–104, 77.

[28] David Arnold, *Colonizing the Body: State Medicine and Epidemic Disease in Nineteenth Century India* (Berkeley, 1993), 5; David Arnold, *Science, Technology and Medicine in Colonial India, The New Cambridge History of India*, iii–5, (Cambridge, 2000), 68–71.

[29] Alexander Duff, *The Church of Scotland's India Mission. An Address delivered before the General Assembly*, 26 May 1835 (London, 1835), 6, 12.

O'Shaughnessy had a similar attitude is unlikely, since there is no evidence of any strong religious bias in his life and certainly he never expressed such opinions in writing. His main preoccupation was science- and botanical-based medicines, topics about which he frequently stressed his indebtedness to native doctors, and in his research, he applied the knowledge gained from them about botanical medicines and, of course, he also studied historical documents in his quest for new therapies. It is doubtful if his work was other than genuinely scientific.

On first arriving in India, O'Shaughnessy was assigned briefly to Fort William College in Calcutta, an institution founded in 1800 with the intention of educating Company employees in Indian languages, although study at the College was principally for civilian officers rather than for surgeons. The origin of the College with its emphasis on language proficiency was based on the work of a Scottish East India Company physician, John Borthwick Gilchrist (1759–1841), who began to study Urdu and in 1787 was granted a year's leave from his medical duties to work on an Urdu-language dictionary. In fact, he never returned to the medical service, instead moving to Calcutta where he wrote several books to help young men arriving to join the Company's service and began teaching them Indian languages, very successfully.[30] As a result of his work the governor-general, Marquis Wellesley, and the Company officers agreed on 10 April 1801 to found an institution to be known as The College of Fort William with the intention of teaching new arrivals to the East India Company service native languages.[31]

O'Shaughnessy's interest in traditional Indian medicine may have originated during this secondment to Fort William, where the study of Indian language and culture was the College's raison d'être. Certainly, his three-month attachment to the College would have introduced him to Indian languages and to Indian culture and literature. It is quite possible that he was more receptive to Indian languages and traditions because of his Irish origin and upbringing in County Clare where two languages, Gaelic and English, and two cultures, Irish and Anglo-Irish, existed side by side. His own parental background with a Catholic father from an ancient Irish clan and a mother who was Protestant and from an Anglo-Irish family must have given him a tolerance and an acceptance of different ideologies and cultures. It is certain that he did speak Gaelic and of

---

[30] Kathleen Prior, 'Gilchrist, John Borthwick 1759–1841 philologist', *Oxford Dictionary of National Biography* (2023). In India, Gilchrist had a long relationship with an Indian Moslem woman and had two sons and five daughters with her, some of whom he took back to Edinburgh when he left the service, before eventually settling in London in 1816.

[31] Katharine S. Diehl, 'The College of Fort William', *Libraries and Culture* (2001), https://web.archive.org/web/20070430072337/http://www.gslis.utexas.edu/~landc/bookplates/13_4_FtWilliam.htm (accessed 8 July 2024).

course, although educated in English, the language and elements of Gaelic culture would have been familiar to him.[32] As a medical student in Edinburgh, not only would he have heard both Scots and Gaelic spoken by fellow students but also from patients and in and around the city streets and of course he married in an Edinburgh Presbyterian church a Scots woman who spoke both Scots and English. These experiences in Ireland and in Scotland may have made him more tolerant of Indian culture and native medicine but undoubtedly the greatest element in his pursuit of Indian therapies was his own intellectual curiosity. Moreover, as Crosbie points out 'recent studies typical of the "new imperial history" have demonstrated that Irish, English, Scottish, and Welsh personnel in fact viewed the empire in different ways and interacted with indigenous people and culture accordingly... Ireland was thus both "imperial" and "colonial" at the same time, "colonizer" but also "colonized"'. As John Mackenzie puts it, 'members of each ethnicity interacted with empire, and its indigenous peoples, in different ways'.[33] Whether O'Shaughnessy's upbringing in two worlds and his brief Scottish experience caused him to be more tolerant of Indian people and culture is impossible to answer but it is probable that he was more sympathetic.

Within a few months of his arrival in Bengal he found time not only to join the Asiatic Society but also to give a paper to the Society in March 1834, meanwhile carrying out his official duties as an assistant surgeon in the East India Company army. In early 1834, he was in medical charge of Civil Stations, first at Gyah and then Cuttack, before serving with the artillery at Dum Dum and later with the 72nd Bengal Native Infantry. In December 1834, he was sent to take medical charge of a detachment of the 72nd Native Infantry, ordered to the Upper Provinces on escort duty with the camp of the Honourable the Governor of Agra.[34] Early in 1835 he was with the 10th Regiment Bengal Light Cavalry before he was sent to assist the Opium Agent at Behar, until in August 1835 he was appointed Professor of Chemistry in the Medical College of Calcutta.[35] Although he would have been very involved in surgical duties when attached to these regiments, the fact that from day to day he was working

[32] When he was appointed foreign secretary of the Medical and Physical Society of Calcutta, he jokingly remarked that his only qualification to be described as foreign was his knowledge of Gaelic.

[33] Barry Crosbie, 'Ireland, Colonial Science, and the Geographical Construction of British Rule in India, 1820–1870', *The Historical Journal*, 52, 4 (2009), 963–987, 964, 966; see also John M Mackenzie, 'Irish, Scottish, Welsh and English Worlds? A Four Nation Approach to the History of the British Empire', *History Compass*, 6, (2008), 1244–1263.

[34] 'Military Appointments and Promotions', *Asiatic Journal and Monthly Register for British and Foreign India, China and Australia*, 17 May–August 1835, 127.

[35] M. Adams, *Memoir of Surgeon-Major Sir William O'Shaughnessy Brooke in*

70   AN INNOVATIVE PHYSICIAN AND SCIENTIST

with Indian doctors who had been trained in Ayurvedic medicine must have stimulated his interest in evaluating native plants and their possible therapeutic uses. Thus, in only eighteen months between his arrival in Calcutta as a very new young assistant surgeon he had become the first Professor of Chemistry in the new College of Medicine and by 1839 he was ready to publish a detailed and well-researched monograph on *Cannabis indica.*

It is hard to believe that O'Shaughnessy arrived in Bengal little more than a year after his innovative research on the chemical pathology of cholera had been carried out in the north of England and published in *The Lancet* in 1832. He was now on his way to where the cholera epidemic had originated, little more than one hundred miles from Calcutta, where initially he was to be based. It was evident that his reputation as an expert chemical analyst preceded him, as extracts from the Proceedings of the Asiatic Society reveal, published in the *Indian Journal of Medical Science* in March 1834. The committee wrote:

> The wide field of vegetable chemistry has been hitherto nearly untrodden in India; and yet there is no country where it offers a richer harvest of curious and novel results. We hope Dr O'Shaughnessy's talents once directed to the subject, will be fixed on this difficult branch of chemical analysis. He has already acquired in England the peculiar skill and experience in recognising and separating the numerous and complicated principles of which organic substances are composed...[36]

His reputation and research had undoubtedly made an impression on the committee of the Asiatic Society, whose members included Sir Charles Metcalfe, who had been elected one of the Vice-Presidents that same evening and who will figure in the story later. Also present were John Tytler, Charles E. Trevelyan and Dr Nathaniel Wallich, all of them distinguished in their own fields and the latter important to O'Shaughnessy's later career in India.[37] The Asiatic Society was founded by thirty British residents of Calcutta in

---

Connection with the Early History of the Telegraph in India (facsimile edition by Kessinger Legacy Reprints of edition Simla, 1889), 5.

[36] 'Report of the Proceedings of the Asiatic Society on 30 January 1834', *Journal of the Asiatic Society of Bengal*, 145, 146.

[37] John Tytler (1787–1837), a Scottish surgeon and leading Orientalist, was a friend and colleague of both Horace Hayman Wilson and O'Shaughnessy; for a full appraisal of his career, see Michael J. Whitfield, 'Dr John Tytler (1787–1837), Superintendent of the Native Medical Institution, Calcutta', *Journal of Medical Biography*, 28, 4 (2021), 84–189; C.E. Trevelyan (1807–1886), later Sir Charles Trevelyan, Bart, arrived in Calcutta in 1831 having previously been secretary to Sir Charles Metcalfe in Delhi from 1827; G. Boase and D. Washbrook, 'Trevelyan, Sir Charles Edward, First Baronet (1807–1886), Administrator in India', *Oxford Dictionary of National Biography*, www-oxforddnb-com.ezproxy.is.ed.ac.uk/view/10.1093/

1784 with a somewhat ambitious, even grandiose statement of intent: 'The bounds of investigations will be the geographical limits of Asia, and within these limits its enquiries will be extended to whatever is performed by man or produced by nature.' For most of its early existence only Europeans were permitted to join the Society but in 1829 several distinguished Indian scholars became members, and a 'Native Secretary' was appointed in 1832. By the time O'Shaughnessy was elected a member, the Society's aims had been greatly modified with its journal recording its purpose as being both 'antiquarian and linguistic encompassing also the natural sciences and medicine' with the latter discipline much more prominent than had previously been the case. This marked a change from a previously antiquarian bias to one which was influenced by the creation of the Calcutta Medical and Physical Society in 1823, whose meetings took place at the Asiatic Society and were reported in the latter's journal, emphasising the interest shown locally in new medical developments in the West:

> the expediency of making some of the Society's [i.e. the Medical and Physical] publications convey early intelligence of important discoveries in medicine, and of the progress of medical science in general, without any expense to the members, has been already twice brought to the notice of the Society, and some steps taken with a view of altering the form, and very much increasing the size of the Monthly Circular; for the purpose of comprehending an account of the progress of medical science in other parts of the world, so arranged as to admit of being bound up in a volume at the end of each year. The Society's annual publications would thus comprise one volume, principally composed of the discoveries in every branch of the profession in other parts of the world.[38]

It is evident from the records of the Asiatic Society that O'Shaughnessy wasted no time in beginning scientific research: in March 1834, a mere four months after his arrival in Calcutta, having spent some obligatory introductory time in the college at Fort William, he gave a talk to the society members on the properties of an edible moss to which he gave the name *fucus amylaceous*. The 'edible value' of this moss, he suggested, might be used in a convalescent jelly to aid recovery from various disorders both on land and at sea.

---

ref:odnb/9780198614128.001.0001/odnb-9780198614128-e-27716 (accessed 30 October 2023). Metcalfe and Wallich will be discussed later in more detail.

[38] 'Report of the Medical and Physical Society of Calcutta', *Journal of the Asiatic Society of Bengal*, 1 (1832), 225–255. The Medical and Physical Society of Calcutta was a society of British officials, largely physicians, formed on 1 March 1823. The society, which met at the premises of the Asiatic Society, published a quarterly journal with articles on diseases prevailing in India and their links with environment and sanitation.

72   AN INNOVATIVE PHYSICIAN AND SCIENTIST

Its presence locally in abundance may have gone some way to satisfy the East India Company that something was being done in response to their call for economies in the provision of drugs with some research into the potential of local plants.[39] Soon after his work on edible moss, he went on to examine the properties of several more native plants, addressing the Medical and Physical Society on the chemical properties of the trees, neem and rohema, which were reputed to have similar therapeutic effects to quinine, an essential medication in tropical climates where malaria, an intermittent fever, was common and frequently fatal.[40] The high incidence of malaria in India and its treatment with quinine, which had to be imported from South America where Britain had little colonial presence, was highly expensive and there was no guarantee of continued supply.

At a meeting on 1 March 1834 of the Medical and Physical Society, Dr O'Shaughnessy read an extract from an unspecified medico-chirurgical journal in which was an article informing the medical profession that the Peruvian government was about to suspend for five years the export of cinchona bark, the source of quinine, in order to preserve the country's cinchona trees whose numbers were being depleted by excessive destruction. The bark of the cinchona, a large shrub, was the source of the quinine which was the specific treatment of remittent and intermittent fevers, one of which was malaria, a disease which was an economic and manpower drain on the expanding empire from repeated illness and an all too high mortality. It was not coincidental that the search for a quinine substitute was one of the early research projects O'Shaughnessy carried out in Calcutta: first into neem and rohema, and secondly, narcotine. These were to replace quinine but his major work on cannabis was quite different; it was a new medicine. It might almost appear as if O'Shaughnessy had been sent to Bengal to research and develop local medicines.

The neem tree (*azadirachta indica A*. Juss), the Indian lilac, was used extensively in both Ayurvedic and Unani medicine for a variety of ailments. The leaves, bark, fruit, flowers, oil and gum yield several biologically active chemical constituents, including many with antioxidant, antimalarial and anticarcinogenic activity and was used in the nineteenth-century as a fever treatment.[41] The rohema, otherwise *rohana* in Sanskrit, is a tree found in

[39] W.B. O'Shaughnessy, 'On the Composition and Properties of the Fucus Amylaceous', *Indian Journal of Medical Science*, 1, 1 (1834). This research was carried out and written in February 1834 when he was attached to the Calcutta General Hospital.

[40] *Asiatic Journal and Monthly Register for British India*, 15:59, November 1834, recorded that Dr O'Shaughnessy had addressed the Medical and Physical Society on 1 March 1834.

[41] G. Brahmachari, 'Neem-An Omnipotent Plant: A Retrospection', *ChemBioChem*, 5 (2004), 408–521.

India which was also thought to have antifever properties, hence the reason for O'Shaughnessy's interest, and of course he was hoping to find a cinchona bark and quinine substitute from both trees. These were not the only avenues he was exploring, for he later described to the Council the results of treating patients suffering from remittent or intermittent fever, malaria, with narcotine instead of quinine.[42]

His research into narcotine, readily available in India, derived from opium of which there was locally a massive supply, showed that it was a possible substitute for quinine in the treatment of these fevers, results which he reported later in *The Lancet*: 'On 4 August 1838 at a meeting of the Medical Society of Calcutta, Dr O'Shaughnessy laid before the Society the details of thirty-two cases of remittent and intermittent fevers treated by narcotine as a substitute for quinine and of which thirty-one were cured.' He included reports from several other physicians and surgeons who had treated similarly affected patients successfully, including one from Mr Green, civil surgeon at Howrah, who had treated sixteen patients and considered narcotine a 'more powerful antiperiodic than quinine'. Interestingly, included in the report were cases from the Native Hospital in Howrah, cases treated by Pandit Madhusudan Gupta and presumably treated with narcotine. Gupta later became an important source of information and advice when O'Shaughnessy became involved in his research on cannabis.[43] The absence of a confirmatory diagnosis in all the cases, an essential part of a drug clinical trial in the present day, perhaps raises questions as to the type of fever being treated and since narcotine has good anti-inflammatory properties, it is possible that this was the reason for successful outcomes in conditions which may well not have been malaria.[44] Of course, in the nineteenth century confirmation by blood tests was not possible.

[42] 'Report of the Medical and Physical Society of Calcutta', *Calcutta Monthly Journal* (1838), 387.

[43] According to Jayanta Bhattacharya, 'From the Inception to the First Dissection Calcutta Medical College 1836', *Doctors' Dialogue* (2023), 2; Pandit Madhusudan Gupta (1800–1856), a Baidya Brahmin linguist and Ayurvedic physician, professor at the Sanskrit College, belonged to a traditional Bengali physician caste. He transferred to the new Calcutta Medical College where it is thought that he supervised the first anatomical dissection.

[44] *Quarterly Journal of the Medical and Physical Society of Calcutta* (1837–1838). At this time the joint editors were Professors Goodeve and O'Shaughnessy; William Brooke O'Shaughnessy, 'Analysis of the Edible Moss of the Eastern Archipelago', *Journal of the Asian Society of Bengal*; William Brooke O'Shaughnessy, 'On the Composition and Properties of the Fucus Amylaceous', *Indian Journal of Medical Science*, 1, 1 (1834); W.B. O'Shaughnessy, 'On Narcotine as a Substitute for Quinine in Intermittent Fever', *The Lancet* (20 July 1839), 606–607, 607 quoting the *Calcutta Quarterly Journal and the Indian Journal of Medical Science.*

74 AN INNOVATIVE PHYSICIAN AND SCIENTIST

It was not only in Bengal that there was a growing interest in his pharma-cological research, for in England *The Lancet* also published articles in 1839 on the preparation of narcotine, on its properties and the results of its use as a substitute for quinine in the treatment of intermittent fever, citing O'Shaughnessy. He had used what he described as a 'simple and economical' method to isolate narcotine from opium, the latter was of course being in plentiful supply in Bengal, observing that the cost of narcotine produced by his process would be one quarter that of quinine. A second paper reported on the successful use of narcotine in the treatment of fifty-nine cases of remittent and intermittent fevers in which all but two of the cases treated were cured.[45]

Many scholars have examined the role of medicine in the empire. One scholar, Mark Harrison, in an essay on British medical topography in India, rightly pointed out that:

> After several decades of scholarship on science and empire, it is now largely accepted that disciplines such as medicine and geography played a crucial role in imperial expansion. On a purely technical level these disciplines were important 'tools of empire', enabling colonizers to map their new domains...[46]

Although the author was addressing mainly geographical and topographical questions that were of importance to the British in India, there is some evidence that the 'science of medicine' as much as topographical concerns had become key, and especially the science of pharmacology. The development of new drugs from native plants arose in part from the East India Company's desire for economies. Because the Company hoped to save money on the manufacture and transport of drugs to India, they encouraged their medical staff to explore Indian flora and the history of their use in Ayurvedic medicine. Fortunately, there were many East India Company surgeons who had a broad scientific education and were expert botanists, notably men who had studied in the Edinburgh medical school where botany remained an essential part of a medical education.[47] William Brooke O'Shaughnessy is a perfect example of a medical scientist with a training in botany; he was also a chemist and a pharmacologist with an astonishing breadth of interest and knowledge as will

---

[45] *The Lancet*, 32 (20 July 1839), 606–607.

[46] Mark Harrison, 'Differences of Degree: Representation of India in British Medical Topography 1820–c.1870', *Medical History Supplement*, 20 (2000), 51–69, 1.

[47] Dr Henry Noltie of the Royal Botanic Gardens Edinburgh has written extensively on Scottish educated surgeons who were expert botanists, men such as Robert Wight, Hugh Cleghorn and others: Henry Noltie, *The Botany of Robert Wight* (Liechtenstein, c. 2005); Robert Wight (1796–1872) was an Edinburgh MD who served as an assistant surgeon with the East India Company.

be revealed when his remarkable work, first editing almost single-handedly the *Bengal Pharmacopoeia* and his work on cannabis is examined.

This section, therefore, will seek to examine the reasons why O'Shaughnessy became interested in the plant *Cannabis indica* and its potential use both in India and in Europe as a treatment for several medical conditions and importantly for the relief of pain. His interest contrasted with that of many of his fellow East India Company surgeons, few of whom had exhibited much interest in Indian drug treatments; of course, it is likely that very few of them had his enquiring mind or his training in botany and none of them established themselves as scientists in Bengal so rapidly and so enthusiastically.

It was the work of men like George Playfair who in his translation of a work of Indian *materia medica* found the following assessment of cannabis:

> *Cannabis Sativa*, a name for *Kainib*, called also *Bidjia*; it is pungent, bitter, hot, light, and astringent; it promotes appetite, cures disorders of phlegm, produces idiotism; is the cause of foolish speech and conduct, or in other words, it intoxicates; if used in excess it produces fever, and it increases all the deleterious effects of poison. Take of *Bidjia* 64 tolahs, when the sun is in the division *Sirtaam,* white sugar 32 tolahs, and pure honey 16 tolahs, cow's ghee 34 tolahs. First fry the Bidjia in the ghee, then add the honey in a boiling state, afterwards the sugar: use this in moderate doses daily, and when it has been used for two months, strength and intelligence will have become increased, and every propensity of youth restored; the eye-sight cleared, and all eruptions of the skin removed; it will prove an exemption from convulsions and debility, and preserve the bowels at all times in a state of order. It will likewise give an additional zest for food.[48]

The appraisal tells of the drug's ability to prevent convulsions, increase appetite and 'preserve the bowels' if taken in moderate doses. As a scholar of the history of cannabis writes:

> O'Shaughnessy listened to local lore, then effected animal studies and human trials to demonstrate the efficacy of cannabis extracts in the frequently fatal diseases of tetanus and cholera, and in providing a more peaceful passage to inevitable demise in rabies. Soon *Cannabis indica* and extracts were exported to Great Britain and an enthusiastic bout of experimentation extended to Europe and America.[49]

The results of his research were published first in the *Journal of the Asiatic Society of Bengal* in October 1839, by which time the author was joint secretary of the Asiatic Society and four years into his tenure as Professor of Chemistry

---

[48] Playfair, *The Taleef Sereef* (Calcutta, 1833), number 248.
[49] Russo, 'History of Cannabis', 4, 5.

76 AN INNOVATIVE PHYSICIAN AND SCIENTIST

and Materia Medica in the new Calcutta Medical School. In acknowledging his debt for assistance in the historical and geographical uses of cannabis, both as a recreational substance and a medicine, he mentions distinguished members of the Hindu and Muslim communities with whom he consulted. For advice on early Sanskrit writers on the use of cannabis in medicine, he thanked Pandit Gupta and other colleagues for extracts from Hindu and Persian writings on how the drug was used in the past. He wrote, 'Such was the amount of preliminary information before me, by which I was guided in my subsequent attempts to gain more accurate knowledge of the action, powers and possible medicinal applications of this extraordinary agent.' In his introduction he thanked several colleagues who were knowledgeable and had experience in the use of cannabis:

> in the historical and statistical department I owe my cordial thanks for most valuable assistance to the distinguished traveller the Syed Keramut Ali, Mootawulee of the Hooghly Imambarrah, and also the Hakim Mirza Abdul Rhazes of Teheran, who have furnished me with interesting details regarding the consumption of Hemp in Candahar, Cabul, and the countries between the Indus and Herat. The Pundit Modoosudun Goopto has favoured me with notices of the remarks on these drugs in the early Sanskrit authors on Materia Medica; – to the celebrated Kamalakantha Vidyadanka, the Pundit of the Asiatic Society, I have also to record my acknowledgments; – Mr. DaCosta has obligingly supplied me with copious notes from the 'Mukzun-ul-Udwieh,' and other Persian and Hindee systems of Materia Medica. For information relative to the varieties of the drug, and its consumption in Bengal, Mr. McCann, the Deputy Superintendent of Police, deserves my thanks; – and lastly, to Dr. Goodeve, to Mr. Richard O'Shaughnessy, to the late Dr. Bain, to Mr. O'Brien of the Native Hospital, and Nobinchunder Mitter, one of my clinical clerks, I am indebted for the clinical details with which they have enriched the subject.

Clearly, he had consulted widely on how cannabis was used in different parts of Asia and the Middle East, both contemporary and historically, with the first eighteen pages of O'Shaughnessy's article taken up by a detailed, historical review of the many uses of cannabis from the earliest time to the present, and from Europe to Egypt, Persia and Asia. He explained his purpose:

> I first endeavour to present an adequate view of what has been recorded of the early history, the popular uses and employment in medicine of this powerful and valuable substance. I then proceed to notice several experiments which I have instituted on animals, with the view to ascertain its effects on the healthy system; and, lastly, I submit an abstract of the clinical details of the treatment of several patients afflicted with hydrophobia, tetanus, and other convulsive disorders, in which a preparation of Hemp

BENGAL DISPENSATORY AND *CANNABIS INDICA*

was employed with results which seem to me to warrant our anticipating from its more extensive and impartial use no inconsiderable addition to the resources of the physician.

A scholar introducing his review of O'Shaughnessy's paper on the Indian hemp observed that 'the pamphlet contains a detail of facts of a very important kind, which, we doubt not, will cause a great sensation among the members of the profession throughout the world'.[50]

Section Five of his monograph described his experiments on animals to ascertain its effects, after which he relates how he used the drug on humans, introducing this section writing:

> the influence of the drug in allaying pain was equally manifest in all the memoirs referred to. As to the evil sequelae so unanimously dwelt on by all writers, these did not appear to me so numerous, so immediate or so formidable as many which may be clearly traced to over-indulgence in other powerful stimulants or narcotics viz. alcohol, opium or tobacco.

His methodology was then to carry out animal experiments which nowadays of course would be considered totally unethical, but he emphasised that in none of these experiments was there any evidence that the dogs used were at any time in pain. Only after he was satisfied that the risk was negligible, did he proceed to human trials using the drug in a variety of medical conditions. He concluded saying that he had given a large supply of gunjah to Mr Squire of Oxford Street, a London pharmacist, who 'promised to make a sufficient quantity of the extract available to any hospital physician or surgeon who may desire to employ the remedy'.[51]

The *British and Foreign Medical Review* of his monograph wrote that 'the profession is under great obligations to Dr O'Shaughnessy for the publication of the important facts detailed in the pamphlet... the pamphlet contains a detail of facts of a very important kind, which we doubt not will cause a great sensation among the members of the profession throughout the world.'[52] The *Review* thought that one of the most important points made by the author was dosage:

---

[50] Anon., 'On the Preparations of the India Hemp, or Gunjah (Cannabis Indica), their Effects on the Animal System in Health and their Utility in the Treatment of Tetanus and other Convulsive Diseases', *British and Foreign Medical Review* (1840), 225–271, 225.

[51] O'Shaughnessy, 'On the Preparations of the Indian Hemp', 347–369, 369.

[52] *British and Foreign Medical Review*, x, (1840), 225–228, 225.

The dose in which the hemp preparations might be administered, constituted, of course, one of the first objects of his inquiry. Ibn Beitar, an Arab physician, had mentioned a direm or forty-eight grains of churrus: 'but this dose seemed to me so enormous, that I deemed it expedient to proceed with much smaller quantities.'[53]

Peter Squire, the eminent London pharmacist who had been given cannabis by O'Shaughnessy, wrote the following on his experience and that of others on the actions of *Cannabis indica*, Indian hemp:

> We are indebted to Dr. O'Shaughnessy for the first introduction of Indian Hemp into this country. He brought over a quantity from India, which the Author converted into extract for him, and distributed amongst a large number of the profession under Dr O'Shaughnessy's directions. Dr. Clendinning used it largely, and his opinion is as follows : – "It acts as a soporific or hypnotic in conciliating sleep; as an anodyne in lulling irritation; as an antispasmodic in checking cough and cramp; as a nervine stimulant in removing languor and anxiety, and raising the pulse and spirits without any drawback or deduction on account of indirect or incidental inconveniences, producing tranquil sleep without causing constipation, nausea or other effect or sign of indigestion, without headache or stupor." More recently, Dr. Russell Reynolds has found it very successful in certain cases of insomnia, neuralgia, and spasm. He says it relieves these derangements of the nervous system, without interfering with any one of the functions of organic life and does not produce the after suffering of misery which follows many opiates. Most valuable in allaying all disturbance of the spinal cord.[54]

Peter Squire recounted the views of one of the physicians using the new drug, Dr John Clendinning (1798–1848), senior physician at the St Marylebone Infirmary, who reported his experience of prescribing cannabis: 'the assertion, that the addition to our materia medica, of any remedy possessing to any considerable extent the virtues without the defects of opium, would be an advantage not easily overrated. Now such an agent I suspect we possess in the Extract. Cannabis Indica.' His intention was to 'determine as nearly as I could, the question, whether the hemp narcotic be in reality possessed of medicinal properties sufficiently energetic and uniform to entitle the drug to admission into our pharmacopeia'. After treating many patients, he was quite certain that cannabis was superior to opium and had fewer unpleasant side effects, writing of its merits:

---

[53] Ibn Beitar (1197–1248) was an Arab botanist, physician and pharmacist.

[54] Peter Squire, *Companion to the latest Edition of the British Pharmacopoeia comparing the strengths of various preparations with those of the London, Edinburgh, Dublin and United States* (London, 1884), 74–75.

BENGAL DISPENSATORY AND CANNABIS INDICA

[Its] effects as a soporific or hypnotic in conciliating sleep; as an anodyne in lulling irritation; as an antispasmodic in checking cough and cramp; and as a nervine stimulant in removing languor and anxiety and raising the pulse and spirits; and that these effects have been observed in both acute and chronic affections, in young and old and male and female.

Dr Michael Donovan (1790–1848), a Dublin apothecary and pharmacist, reported the successful treatment of neuropathic and musculoskeletal pain with *Cannabis indica*, not only prescribed by himself but also by the distinguished physician, Dr Robert Graves (1796–1853), who used it successfully in cases of neuralgia. Donovan was in no doubt as to the value of the drug:

I indulge in the expectation that this powerful agent, when physicians have fully developed its properties, will rank in importance with opium, mercury, antimony, and bark. The public and the Profession owe a deep debt of gratitude to Professor O'Shaughnessy, whose sagacity and researches have brought to light a medicine possessed of a kind of energy which belongs to no other known therapeutic agent, and which is capable of effecting cures hitherto deemed nearly hopeless or altogether impracticable.

It is interesting that Donovan noticed one very important drawback: 'it became obvious how much the plant suffers from age', a characteristic which physicians were becoming increasingly conscious of and making many a little wary of this new medication. He had reservations: 'The difficulty, or rather impossibility, of determining what would be an effective dose for a patient of whom the practitioner has had no experience, with reference to the intensity of the pain and the susceptibility of the patient, has greatly limited the employment of this important medicine.'[55]

Robley Dunglison (1790–1869), Professor of Medicine in Jefferson Medical College, Philadelphia, in an 1843 article on 'new remedies, pharmaceutically and therapeutically considered', discussed the use of *Cannabis indica*, saying that the first person who 'seems to have well tested its properties is Dr O'Shaughnessy', who encouraged by the result of animal testing, felt justified in giving 'the resin of the hemp an extensive trial in cases in which its apparent powers promised the greatest degree of utility'.[56] Dunglison considered that the general effects on humans were encouraging with alleviation of pain noted in most cases and observed that the results from using cannabis in cases of

---

[55] Michael Donovan, 'On the Physical and Medicinal Qualities of Indian Hemp, Cannabis Indica with Observations on the Best Mode of Administration, and Cases Illustrative of its Powers', *Dublin Journal of Science*, 26, 3 (January 1845), 368–462. Donovan grew Indian hemp in order to test its strength in the northern hemisphere.

[56] Robley Dunglison, *New Remedies: Pharmaceutically and Therapeutically Considered* (Philadelphia, PA, 1843), 133–136.

tetanus were remarkable where nine patients out of fourteen recovered. He quoted O'Shaughnessy, who had observed 'to me they seem unequivocally to shew, that when given boldly and in large doses, the resin of Hemp is capable of arresting effectually the progress of this formidable disease, and in a large proportion of cases of effecting a perfect cure'.

Alexander Christison (1828–1918), the eldest son of Professor Robert Christison, in a prize essay on Indian Hemp which was based largely on his doctoral thesis, wrote that 'the expectations held out by him [O'Shaughnessy] have not yet been fully realised. One explanation of this appears to be, that many have used spurious or ill-made preparations, and others have not made sufficient allowance for the influence of constitutional peculiarities in affecting its action.' Christison then described an experiment carried out in the Edinburgh Botanic Gardens using seeds sown in the gardens' 'stove-house' to ascertain whether the Indian plant secreted a much larger proportion of resin than is the case with the European grown plant. It was concluded that this was undoubtedly the case and later he observed that 'the activity of the preparations of hemp depends on the presence of the resinous varnish on the leaves'. This was astute, confirming a problem of which many physicians were becoming increasingly aware.

His father, Robert Christison, he recalled, 'administered hemp in many instances', and observed that 'in the generality of cases, hemp had the effect of causing sleep without disturbing the function of the stomach or bowels... And in some cases he [had] found it to succeed where morphia and opium had failed.' Nevertheless, despite such positive outcomes he had reservations, expressing his doubts about the action of Indian hemp when describing the unsuccessful efforts of various chemists and pharmacists to detach the active principle of the plant: 'the discovery of the true active principle, or even of a uniform resin is, however a great desideratum in medicine; for until this object be attained, it is scarcely possible to rely on the action of any preparation of the plant'.[57] In a subsequent article, Christison when investigating the action of cannabis on uterine contractions, concluded that Indian hemp may often 'prove of essential service in promoting uterine contraction in tedious labours but more experience was needed to show how far the effects might be depended on and in which cases hemp was indicated'.[58] His emphasis on discovering the true active principle and a uniform resin were the comments of a toxicologist and a physician who needed more details of the chemical properties of this very new and largely untried drug.

[57] Alexander Christison, 'On the Natural History, Action and Uses of Indian Hemp', *Edinburgh Medical Journal* (July 1851), 26–45, 34.

[58] Alexander Christison, 'On the Natural History, Action and Uses of Indian Hemp', *Monthly Journal of Medical Science* (August 1851), 117–121.

Dr James Adair Lawrie, surgeon to the Glasgow Royal Infirmary, reported on twenty-six patients, all women, some who had been admitted to the Lock Hospital with venereal disease. He pointed out that the Indian hemp 'was given for the purpose of ascertaining its physiological more than its therapeutical properties', an experiment quite unethical by today's standards. He concluded that it was not a valuable addition to narcotic medicines, and it did not succeed where opium had failed.[59] He like his colleagues compared the effectiveness of cannabis to opium. In 1869, Dr Douglas addressed the Medico-Chirurgical Society of Edinburgh on the 'Use of Indian Hemp in Chorea', observing that cannabis 'has an important application, not only to distressing and dangerous cases of chorea, but even to slight and ordinary cases, as well as to cases of other spasmodic disease...' He pointed out that small and frequent doses proved both safe and effective and that increasing the frequency of the dose rather its amount bestowed a great advantage.[60]

Peter Squire had earlier made brief reference to having given samples of cannabis to Dr J. Russell Reynolds (1828–1896), a highly influential London physician and neurologist. Reynolds later wrote in 1890 a paper in *The Lancet* on the uses of *Cannabis indica* and its toxic effects. Reynolds referred in a paper from 1848 to an occasion where a physician had made the comment that '[he] had expected much but it was so uncertain in its action, and its exhibition was sometimes accompanied by such distressing toxic effects that he had discontinued its use'. Reynolds explained how he had obtained from Mr Peter Squire, the London pharmacist, some excellent samples of cannabis from India; apparently for some years Squire had been unable to obtain a good sample of the drug from India or even obtain a plant specimen but now he was in possession of excellent samples and alcoholic extracts which he had passed on to Dr Reynolds hoping he would give the drug a trial. Reynolds concluded that 'with this general result, that Indian hemp, when *pure* and administered carefully, is one of the most valuable medicines we possess' (author's emphasis). He found the most helpful results in mental conditions such as senile insomnia and certain forms of delirium, in neuralgic conditions and migraine, and in muscular clonic spasms. The toxic effects he attributed

[59] James Adair Lawrie, 'Cases Illustrating Some of the Effects of Indian Hemp', *Monthly Journal of Medical Science*, 4 (November 1844), 939–948. Lock hospitals were found in all major cities and in the colonies often as military establishments, where troops and prostitutes could be treated for venereal disease. The name Lock is thought to be derived from the French word *loques*, meaning bandages rather than a place of detention although that was part of the purpose.

[60] Dr Douglas, 'On the Use of Indian Hemp in Chorea', *Edinburgh Medical Journal*, 14, 9 (1869), 777–784; the author was probably Andrew Halliday Douglas (1819–1908).

82 AN INNOVATIVE PHYSICIAN AND SCIENTIST

to the great variations in the strength of the drug, explaining that hemp grown during different seasons and in different places frequently varies in the amount it contains of the therapeutic agent. Moreover, he commented that humans vary greatly in their responses to drugs and there are inevitably idiosyncrasies.[61]

It is no coincidence that so many physicians wrote of their disappointment in the results they obtained from prescribing cannabis. As early as February 1843, when speaking to the Royal Medico-Botanical Society, O'Shaughnessy himself made it very clear that cannabis plants deteriorated over time. The Society, although short-lived, was founded with the aim of investigating, by means of communications, lectures, and experiments, the medicinal properties of plants, of promoting the study of the vegetable materia medica of all countries, and of cultivating medical plants.[62] He explained to the Society that this deterioration may have been the cause of the negative reaction from Dr Jonathan Pereira as to the efficacy of cannabis after he had tested the plants sent to him in 1839. Later, Pereira recorded the treatment of a case of tetanus at the London Hospital, 'carefully watched by Dr O'Shaughnessy when the resinous extract was given in increasing doses' with resulting stupor and the cessation of spasms. The general effects of hemp on the human he considered to be alleviation of pain, remarkable increase of appetite, unequivocal aphrodisia and general mental cheerfulness; its main uses were hypnotic, anodyne and antispasmodic.[63]

Cannabis became a frequently used medication during the remainder of the nineteenth and the early twentieth century, but several factors combined to end the perception of the drug as a valuable addition to the pharmacopeia both in the USA and UK. The unpredictability of the strength of the resin obtained from the hemp plant did not sit well with practitioners who were understandably becoming accustomed to accuracy in prescription and dosage. To add to this problem the addictive properties of morphine, a drug that had become too freely available in the USA, had become a cause of official concern and cannabis was now regarded with suspicion because of the rapid increase in those who used it recreationally. That a recreational drug could at

---

[61] John Russell Reynolds, 'On the Therapeutical Uses and Toxic Effects of Indian Hemp', *The Lancet* (22 March 1890), 637–638; Sir John Russell Reynolds was an eminent physician in the Victorian era who held the presidencies of the Royal College of Physicians of London, and of the British Medical Association. He was physician to Queen Victoria.

[62] *Provincial Medical and Surgical Journal* (1843), 1–5: 436, in which is reported a meeting of the Royal Medico-Botanical Society addressed by O'Shaughnessy on 22 February 1843.

[63] Jonathan Pereira, *The Elements of Materia and Therapeutics*, 4th edn (London, 1851), 371–372, 390. This edition was edited by Alfred Swaine Taylor and George Oliver Rees; anodyne means relief of pain.

the same time be a useful medicine was unacceptable. The authors of a book on the medicinal uses of cannabis in their preface wrote of how 'cannabis moved from being a frequently prescribed item for a variety of therapeutic conditions through a period of increasing opposition to it use because for its potential for abuse to the point where its use was completely withdrawn in the mid twentieth century'.[64] Cannabis was removed from the UK pharmacopoeia in 1932 and from the USA pharmacopoeia in 1941.

Professor Mills in *Cannabis Britannica* quotes several eminent British authorities who were convinced of the therapeutic potential of cannabis, including in 1899 Professor W.E. Dixon, FRS, a physician and pharmacologist who argued that it was 'an exceedingly useful therapeutic agent'. Dixon served later on the Rolleston Committee of 1924 appointed to look into the abuse of drugs such as morphine and heroin, but cannabis was excluded from their deliberations as at that time, quite properly, it was not considered to have the same addictive properties as the opiates.[65] Mills questions O'Shaughnessy's motives in promoting cannabis so vigorously, suggesting that 'his excitement about cannabis may not simply have had scientific origins'. He wondered if 'an ambitious and entrepreneurial scientist' was looking for financial gain from the promotion of cannabis in the West.[66] It is true that the fortunes made by companies through selling opium may have persuaded O'Shaughnessy that there was money to be made from cannabis and of course he knew first hand from his time as assistant to an opium agent only too well the fortunes being made by those selling opium to China. Many men who joined the East India Company service in whatever capacity had hopes of making their fortune and many undoubtedly did with the example of Joseph Hume, referred to earlier, perhaps in O'Shaughnessy's mind. Whether financial gain was his main motive is questionable although some of his claims for the drug could be construed as exaggerated. It is worth noting that Mills does not accuse O'Shaughnessy of publishing false information, merely suggesting that he hoped that in time he may have benefited financially from the discovery. His earlier work on the chemistry of cholera brought him no financial benefit and little professional fame or advancement.

Nevertheless, his advocacy for its use in various conditions was tempered by his careful advice on dosage and he made the point that relief of pain was 'manifest in all the memoirs referred to'. O'Shaughnessy's concluding remarks include the following:

---

[64] Geoffrey Guy, Brian Whittle and Philip Robson, *The Medicinal Uses of Cannabis and Cannabinoids* (London, 2004), xiii.

[65] James H. Mills, *Cannabis Britannica. Empire, Trade and Prohibition 1800–1928* (Oxford, 2005), 209.

[66] Mills, *Cannabis Britannica*, 45.

84 AN INNOVATIVE PHYSICIAN AND SCIENTIST

I deem it my duty to publish it without any avoidable delay, in order that the most extensive and the speediest trial may be given to the proposed remedy. I repeat what I have already stated in a previous paper – that were mere reputation my object, I would let years pass by, and hundreds of cases accumulate before publication. But the object I have proposed to myself in these inquiries is of a very different kind. To gather together a few strong facts, to ascertain the limits which cannot be passed without danger, and then pointing out these to the profession to leave them to prosecute and decide on the subject of discussion, such seems to me the fittest mode of attempting to explore the medicinal resources which an untried materia medica may contain.[67]

The limitations of cannabis in pain control became increasingly obvious as hypodermic syringe use became common in the early years of the twentieth century, although an early syringe had been in use from the middle of the nineteenth. Opium-based analgesics were water soluble and could be injected with immediate relief of pain; this was not possible with cannabis which was only soluble in alcohol and unable to be injected. As Alice Mead points out, the water-soluble opiates were injectable 'affording extremely rapid analgesic response' unlike cannabis which was 'primarily administered orally, its onset of action was slow and patient response varied considerably'.[68]

There are contemporary physicians who are opposed to cannabis for various reasons, reasons that are not always logical, and others who are enthusiasts regretting the abandonment of a very useful medication. Dr Lester Grinspoon, a distinguished academic psychiatrist, wrote: 'The government's commitment to gross exaggeration of the harmfulness of cannabis has made it necessary to deny the drug's medical usefulness in the face of overwhelming evidence.' The writer was referring to the USA in 1994 but the theory remains valid for many countries.[69]

Tod Hiro Mikuriya (1933–2007), an American psychiatrist, and author of *Marijuana Medical Papers:1839–1972*, was known as an outspoken advocate for the legalisation and use of cannabis for medical purposes, and is regarded by some as the grandfather of the medical cannabis movement in the United States.[70] In 1969, he wrote:

[67] O'Shaughnessy, 'On the Preparations of the Indian Hemp', 363–369, 368.

[68] Alice Mead, 'International Control of Cannabis. Changing Attitudes', in Geoffrey Guy, Brian Whittle and Philip Robson (eds), *The Medicinal Uses of Cannabis and Cannabinoids* (London, 2004), 369–426, 371.

[69] Lester Grinspoon, *Marihuana Reconsidered* (Cambridge, MA, 1994), 11.

[70] Tod H. Mikuriya, 'Marijuana in Medicine: Past, Present and Future', *Calif. Med.*, 110, 1 (1969), 34–40. PMID: 4883504; PMCID: PMC1503422 (Oakland, CA, 1973).

Cannabis indica, Cannabis sativa, Cannabis americanus, Indian hemp and marijuana (or marihuana) all refer to the same plant. Cannabis is used throughout the world for diverse purposes and has a long history characterized by usefulness, euphoria or evil, depending on one's point of view. To the agriculturist cannabis is a fibre crop; to the physician of a century ago it was a valuable medicine; to the physician of today it is an enigma.

He went on to say, 'the therapeutic use of cannabis was introduced into Western medicine in 1839, in a forty-page article by W.B. O'Shaughnessy, a 30-year-old physician serving with the British in India'. His cannabis clinical trial was truly remarkable for its detailed evaluation of its safety and its qualities at a time, for example, when as was earlier discussed the unscientific prescription of totally untried drugs in the treatment of cholera was being advised by distinguished physicians as the cure-all with no evidence whatsoever to justify such recommendations. It is questionable if any other drug had hitherto been subjected to rigorous testing in this way. Yet today cannabis remains an enigma.

4

Medical Furlough in London and the Royal Society

This chapter covers roughly a decade of what was to be a time of considerable change in O'Shaughnessy's life and career. It will seek to assess the effect of the loss of his position through ill health as Professor of Chemistry and Natural Philosophy in the Calcutta Medical College, an appointment which established him in Bengal as a scholar and innovator not merely an assistant surgeon in the East India Company, his official designation. When he was forced in late 1841 to return to England on medical furlough, this might well have signalled the end of his career, perhaps recalling the disappointment of his failure to succeed in London ten years previously. He was now thirty-three years of age, married for the second time with four daughters to bring up, apparently at the height of his powers with scholarly publications in several scientific disciplines and this setback might well have been devastating. Moreover, another aspect to be considered is that the papers he had written, based on research carried out in the colonies, albeit a major colony, 'the jewel in the crown', would have been considered back in England to be of much less value than work produced in the homeland. This mindset, which was prevalent although unspoken, could have been ruinous to his future prospects, although he did have the friendship with Wakley and through him an opportunity to publish his papers in *The Lancet.* Fortunately, as will be shown, there were many scholars with influence in Britain who were familiar with his work and were sufficiently impressed to support his candidature for the Royal Society.

As the chapter will outline, his sick leave saw him rapidly restored to health, suggesting that overwork was at least partly responsible, although the toll amongst surgeons from tropical diseases and an unaccustomed climate was considerable. O'Shaughnessy was heavily involved teaching chemistry and *materia medica* in the new Calcutta Medical College but with this had come the additional burden of setting up the department, preparing materials for his students, including personally making much of the glass equipment needed for practical chemistry, and finally writing a text on chemistry, a huge workload. At the same time, he was chemical examiner to the government of Bengal, assistant to the assay master of the Bengal mint, secretary of the Asian Society and latterly the single-handed editor of the *Bengal Dispensatory and Pharmacopoeia*, all these activities in addition to his research in photography, telegraphy, and the potential use of native plants in therapeutics.

88 AN INNOVATIVE PHYSICIAN AND SCIENTIST

Thus, until his formal involvement in supervising and establishing the Indian national system of electric telegraphy in the 1850s, the decade of the forties was very much a time of transition for O'Shaughnessy, when his scientific interests and the nature of his work changed or were forced by circumstance to change. From the time of his arrival in Bengal in 1833 where cholera was now endemic, as far as can be ascertained he showed little or no interest in its chemical pathology or its treatment and this did not alter in the 1840s; he did early on as a professor refer to cholera in England but made no mention of his own role. His photographic experiments and research seemed to quickly cease once he had achieved success in producing photographic images and any further trials in electric telegraphy were unnecessary, for he had demonstrated that his systems worked; his main preoccupation at the outset of this transitional period was in exploring the therapeutic potential of native Indian flora, leading to his remarkable clinical trials on the medicinal uses of *Cannabis indica*. This interest in pharmacology was a natural extension of his expertise in chemical analysis but to a certain extent was also driven by the East India Company's need to economise on drugs which were made in the West, then transported to India at considerable expense. For a company whose existence depended on generating profits with which to reward its investors, economies had to be made in a country where the incidence of tropical diseases was high, malaria a scourge, its prevention and treatment costly and supplies of quinine, mainly from South America, uncertain. This need for economies driven by the East India Company's push for self-sufficiency forced O'Shaughnessy to became more involved in *materia medica* and pharmacology. His remarkable dedication and skill in editing the *Bengal Dispensatory* and *Bengal Pharmacopoeia*, published first in 1841, revealed a single-mindedness towards finishing this task, a task made very much more difficult by the unavoidable absence through illness or death of many of his co-editors, to which he referred in his introduction. These volumes rival in quality and diversity the pharmacopoeias produced in the West but also reveal his interest in and awareness of the potential benefits of Ayurvedic therapies, many of which were totally new to Western medicine.[1]

In December 1840, O'Shaughnessy's second wife Margaret and their four daughters left Calcutta for London on the East India Company ship *Reliance*; eleven months later O'Shaughnessy on medical grounds was relieved of his

---

[1]   C. Leslie (ed.), *Asian Medical Systems: A Comparative Study* (Berkeley, 1976) compares the South Asian system which was based on three humours to that of the West with four humours. Chapters on traditional Asian Medicine and Indigenous Medicine in Bengal are especially useful; W.B. O'Shaughnessy, *The Bengal Dispensatory and Pharmacopoeia*. Chiefly compiled from the works of Roxburgh, Wallich, Ainslie, Wight and Arnot, Royle, Pereira, Lindley, Richard, and Fee, and including the results of numerous special experiments (Calcutta, 1841).

professorial duties at the Calcutta Medical College.[2] There is no record as to the nature of his illness but his workload from the time of his arrival in India was considerable, even for a young man. Nevertheless, overwork and stress were not the only possible causes: there were many other factors which could lead to debility and ill-health: the climate of Bengal, the prevalence of and exposure to tropical diseases, and perhaps not least the death of his young first wife, which occurred in 1834 only nine months after arriving in Calcutta. It is hard to comprehend what this loss must have meant to the young doctor in a strange new country, remote from his family in Ireland, with a small daughter, now motherless, to care for.

Whatever the nature of his illness or its cause, it was serious enough to force him to return to England on medical furlough, sailing in November 1841 from Calcutta to London on the *Bangalore*. A long sea voyage must have been arduous, but, on the other hand, the enforced leisure may have been exactly what was needed to aid recovery. It certainly seems to have been restorative for it was not long after arrival in London that he was well enough to become involved in new projects: of course, there is no doubt that election to the Royal Society was the most important of these, but this was not his only scholarly interest or activity during his convalescence. For example, in January 1843 he addressed the Royal Asiatic Society on the improvement of Bengal pottery, how porcelain clay and soap earth found in the hills of Bengal had been used in this endeavour, encouraged by the East India Company; he presented to the Society specimens of the clay and earth which curiously he had brought all the way back from India and strangely there was no mention of cannabis or cannabis plants. It is interesting that this meeting was chaired by the Society's long-serving Director, Professor Horace Hayman Wilson, who was to become both a friend and very influential in O'Shaughnessy's proposed election to the Royal Society.[3]

The following month, O'Shaughnessy was a guest at the Royal Botanical Society of London addressing them on the topic of *Cannabis indica* and its therapeutic uses and at the same meeting he was invited to become a corresponding member of the Society. In his address, which focused on the

---

2   The *Reliance* was later wrecked off Boulogne on a return voyage from China in 1842, having left London in June 1841. The risks of sea voyages were not inconsiderable.

3   Minutes of General Meetings of the Royal Asiatic Society of Great Britain and Ireland, December 1839 to March 1845, 14 January 1843, 120, 121. There is no record that O'Shaughnessy ever became a member of the Society despite a close connection between the London society and the Asiatic Society of Bengal. The Royal Asiatic Society of Great Britain and Ireland was founded in 1823 'for the encouragement of science, literature and the arts in relation to Asia' by H.T. Colebrooke (1765–1837), a notable Sanskrit scholar, who had during his time in Calcutta been, inter alia, president of the Asiatic Society of Calcutta. The connections between the two are clear.

90 AN INNOVATIVE PHYSICIAN AND SCIENTIST

therapeutic value of cannabis, he explained that 'as a paid servant of the government he considered it his duty to examine its [cannabis] properties', meaning its value as a medication, and accordingly he had first of all instituted a series of experiments on animals, prior to treating patients. He went on to say that he had sent Dr Jonathan Pereira specimens of the drug four years previously, but Pereira after conducting trials had not considered that the drug was worthy of further exploration; O'Shaughnessy explained that at that time he was unaware of its deterioration with age, a factor which he considered to be the reason for its failure, the plant having become less potent during the long sea voyage from Calcutta to England. The same month, a London journal published his paper on cannabis.[4] These two societies apart, O'Shaughnessy also belonged to the London Electrical Society, largely a membership of amateur gentlemen with an interest in electricity, especially galvanic transmission, in which of course he had considerable experience. Shortly after attending these society meetings, he was making preparations to leave for Canada.[5]

In March 1843, evidently having recovered from his illness, he sailed from Liverpool to Canada on a Cunard steamer as personal physician to Sir Charles Metcalfe (1785–1846), who had that year been appointed as governor-general of the Province of Canada. Sir Charles required regular medical attention for a malignant facial lesion which ultimately was the cause of his death.[6] Metcalfe had served as temporary governor-general of Bengal in 1835–1836, during which time he had become acquainted with O'Shaughnessy when both were members of the Asiatic Society of Bengal. During this sojourn in Canada, O'Shaughnessy took the opportunity to travel to the USA to visit Joseph Henry and Samuel Morse who were both pioneers of electric telegraphy – more will be written on these two men later. By the middle of 1843, we find him back in Ireland when he wrote from Dublin to H.H. Wilson, seemingly dispirited about his prospects on his return to Bengal. In a letter from 37 Upper Bagot Street, Dublin, dated 5 July 1843, he wrote

---

4   W.B. O'Shaughnessy, 'On the Preparations of the Indian Hemp of Ganja (*Cannabis Indica*)', *Provincial Medical Journal*, 123, 363–369.

5   'Meeting of the Royal Botanical Society, 23 February 1843', *Provincial Medical Journal and Retrospect of the Medical Sciences*, 5, 126, 436–438, 436.

6   John William Kaye (Sir), *The Life and Correspondence of Charles, Lord Metcalfe, Late Governor-General of India, Governor of Jamaica, and Governor-General of Canada; from Unpublished Letters and Journals Preserved by Himself, His Family, and His Friends* (London, 1858). A footnote in this memoir refers to O'Shaughnessy 'as an officer whose great scientific acquirements have since earned him a distinguished reputation'.

MEDICAL FURLOUGH IN LONDON AND THE ROYAL SOCIETY

My Dear Wilson,

I arrived at home two days ago having left our friend Sir Charles Metcalfe in a more satisfactory state than I thought I should ever have seen him in. My return home at a period rather earlier than I originally intended has been caused by Mrs O'Shaughnessy's uncertain health and approaching confinement.

By 1 September he was expecting to return to India 'where however as far as I can see every prospect is clouded and dispiriting'. This pessimism does not seem to have been lessened by his election as a Fellow of Royal Society in March, a major distinction for a mere assistant surgeon in the Bengal medical service but this distinction did not seem to be of moment in Calcutta. He hoped he would get the promised nomination to the Assay Department promised in a letter of 12 August 1842 and he wrote that he hoped Professor Wilson would support his nomination, and if Wilson thought he should go to London and present himself at India House, he would do so without delay.[7]

However, undoubtedly the most significant event during this time on furlough was his election as a Fellow of the Royal Society, the independent scientific society in London which dated from the middle of the seventeenth century, and which rightly could claim to be the most distinguished society of its kind in the world. His election to this eminent body in March 1843 was a remarkable achievement for a man not yet thirty-five years of age, who a mere ten years previously was unable to practise clinical medicine in London because the Royal Colleges of that city considered him unqualified to do so, wishing to maintain a closed shop. The names of the men who supported his nomination were without exception highly distinguished in their own fields often with reputations that have survived to the present day. This next section will record their names in some detail.

O'Shaughnessy's citation for the Fellowship of the Royal Society was impressive:

The Discoverer of the changes caused in the Blood by the malignant cholera – The Author of Memoirs on the state of the blood in cholera – on the detection of poisons – on the laws of the constant Voltaic battery – on the construction of electro-magnetic-machines – on the physiological & medicinal effects of the Indian Hemp – of a dispensatory of materia medica & other works for the use of the students of the medical & Hindu colleges of Bengal – &c &c-. The Inventor or Improver of some forms of Voltaic apparatus &c. Distinguished for his acquaintance with the science

7   BL: H.H. Wilson Collection: MSS. EUR.E.301/7, Letters from O'Shaughnessy, 1843.

92 AN INNOVATIVE PHYSICIAN AND SCIENTIST

of Medicine & Chemistry. Eminent as a Physician and as a promoter of education among the natives of Bengal.[8]

This was indeed a remarkable and varied list of successes for a young physician-surgeon. That such a distinguished scientist as Sir John Herschel Bart (1792–1871) was one of his most enthusiastic supporters must have been a crucial factor in his subsequent election.[9] Herschel was pre-eminent as a scientist in Britain, indeed in Europe, during the first half of the nineteenth century and was a leading light in the Royal Society. It is worth recalling that when in the 1830s a debate had arisen in the Royal Society over the great number of aristocrats with few scientific credentials appointed as Fellows as opposed to bona fide scientists, the council turned to Sir John Herschel to counteract this tendency. He was prevailed upon to stand for the position of President as the following extract from his biography records, indeed a bold, even foolhardy, action considering that his opponent was the brother of the monarch.

> Although known for his judiciousness, tactfulness, and modesty, Herschel in the period around 1830 found himself at the centre of a controversy about science in England. His close friends, Charles Babbage and James South, had been outspoken in claiming not only that science had declined in England, but also that Britain's foremost scientific society, the Royal Society, needed major reform, not least rejection of the practice of electing to membership (and frequently to high office) aristocrats who typically possessed only limited interests in and even less knowledge of science. These criticisms reached the peak of intensity in 1830 when Herschel's friends persuaded him to allow his name to be put forward for the presidency of the Royal Society. His opponent in the election was the Duke of Sussex, the brother of George IV. Although Herschel lost the election 119 to 111, the programme of the reformers was eventually largely adopted.[10]

Herschel's standing in the Royal Society and in the wider scientific world cannot be overstated and this was exemplified by his insistence on academic integrity, electing men of science rather than aristocratic dilettantes; therefore, his willingness to support Dr O'Shaughnessy is even more significant, revealing his admiration for the latter's many original contributions to science and must have influenced his eventual election.

---

[8]   RS, E.C./1843/10. Certificate of Election and Candidature for Fellowship of the Royal Society.

[9]   A coincidence: both Herschel and O'Shaughnessy were married in Edinburgh in March 1829.

[10] Michael J. Crowe, *Oxford Dictionary of National Biography* (2009), https://doi-org.ezproxy.is.ed.ac.uk/10.1093/ref:odnb/13101 (accessed 8 July 2024).

MEDICAL FURLOUGH IN LONDON AND THE ROYAL SOCIETY   93

In Herschel's correspondence, maintained by the Royal Society, there is a letter dated 24 February 1843 from him to Francis Baily (1774–1844), in which he referred to the proposed election of O'Shaughnessy; Bailey was a fellow astronomer, a founder member of the Royal Astronomical Society of which he had been president four times and was on the council of the Royal Society. Herschel expressed the hope that Baily would sign the certificate of proposed election as he himself had done. Francis Baily replied writing from 37 Tavistock Place on 25 February expressing some doubts as to the wisdom of signing the proposal because 'a good deal has been said in the Council respecting the propriety of members of the Council [original emphasis] signing or not signing such certificates'. However, he thought that his presence at the Council meeting will be 'the best assistance I can give him as the qualifications of each candidate are there discussed…'[11],[12] Sir John was clearly of the opinion that O'Shaughnessy's election as a Fellow was merited not only because of his record on cholera and for his qualities as a physician, but also for his abilities in several branches of pure science. He wrote:

> A certain Dr O'Shaughnessy, a very well-informed man, distinguished in Calcutta as a surgeon which he has now left and who has (besides the discovery of the irregular state of the blood in cholera which has acquired him much reputation in his profession) published several clever papers on photography – on Galvanism – and who managed with success the blowing up of some sunken vessels in the Hooghly.[13]

This is a rare and infrequently found reference to O'Shaughnessy's interest in the new science of photography and demands respect coming from Herschel who was at the time himself also experimenting in the technique of daguerreotypes. As a biographer comments, 'nor was chemistry completely abandoned [by Herschel] in fact, in 1819 he published the first and most important of a number of contributions he made to photochemistry. This was his detection that hyposulphite of soda (sodium thiosulphate or "hypo") dissolves silver salts, a major discovery in the prehistory of photography'. Later in his life, his work on photography had a long-lasting influence with his use of sodium thiosulphate recognised as the most useful of all the chemicals for silver-based photographic images. His contributions include experimenting on the light sensitivity of various metals and vegetable dyes, his techniques of making

[11] RS, HS 3. 236. Letter from Francis Baily to Sir John Herschel, 25 February 1843.

[12] RS, HS 3. 235, Letter from Sir John Herschel to Francis Baily, undated February 1843.

[13] RS, HS 3. 234. Letter from Sir John Herschel to Francis Baily, possibly 24 February 1843.

photographs in colour or on glass plates, and his advancing of various terms now standard in photographic science, namely, 'positive', 'negative', 'snap-shot' and 'photographer'.[14]

Daguerre announced his photographic process in August 1839, and it is remarkable in view of distance that by December of that year the *Bombay Times* had published a translation of his work on the daguerreotype. In 1840, the *Asiatic Journal* reported that Professor O'Shaughnessy had demonstrated his 'photographic drawings' to members of the Bengal Asiatic Society in October 1839: 'Before the Meeting broke up Dr. O'Shaughnessy exhibited several Photogenic drawings prepared by himself, and in which a solution of gold was the agent employed.' It is extraordinary that not only was he experimenting in photographic techniques in colour photography so soon after Daguerre had published his results but also that his use of gold was very much his own technical advance.[15] G. Thomas, in the journal *History of Photography*, relates how soon after O'Shaughnessy's arrival in Calcutta he became interested in the work of Daguerre and sought to improve on it by introducing colour to the process. According to the writer:

> He was able to obtain red, purple, and even green tones on the daguerreo-types by using different metals. He announced his results at the meeting of the Asiatic Society of Bengal in August 1839, when he exhibited several 'photogenic drawings prepared by himself in which a solution of gold was the agent employed.'[16]

This early exploration of photographic techniques by both Herschel and O'Shaughnessy reveals how they had much in common, with both men interested in similar areas of scientific research. The support of Herschel in his election cannot be underestimated but the chance of General Pasley calling on O'Shaughnessy to invite him to the grand explosion was providential, all because of Pasley's interest in his destruction of underwater wrecks in the Hooghly. O'Shaughnessy was in the right place at the right time.

[14] For more on Hershel's pioneering research in photography, see Michael R. Peres, Mark Osterman, Grant B. Romer, Nancy M. Stuart PhD, and J. Tomas Lopez, *The Concise Focal Encyclopaedia of Photography: from the first Photo on Paper to the Digital Revolution* (New York, 2007).

[15] 'Proceedings of the Asiatic Society, 2 October 1839', *Journal of the Asiatic Society*, 91 (1839). There is a discrepancy between the dates quoted in the Society journal and that quoted by Thomas.

[16] Dr John Adamson (1809–1870), LRCSEd 1829, an Edinburgh medical student at the same time as O'Shaughnessy, was an early pioneer in photography but his brother Robert Adamson is better known. John produced the first calotype photograph in Scotland in 1842.

The first intimation that O'Shaughnessy was considering such a step as fellowship of the Royal Society can be deduced from his correspondence: on 23 January 1843, he wrote to Sir John Herschel to tell him that General Pasley had called on him that morning, telling him that the 'great explosion' was to come off on Thursday and presumably his visit was to invite him to the event. O'Shaughnessy himself had famously engineered underwater explosions on the River Hooghly, near Calcutta, to destroy wrecks which were hazardous to shipping. The *Bombay Gazette*, impressed by his expertise, in January 1840 reported that Professor O'Shaughnessy had been requested to blow up the remaining part of the *Equitable* and wished to know what the mercantile community would do were not Professor O'Shaughnessy to hand.[17] His expertise and success in underwater explosions were undoubtedly the reasons why Pasley was calling on him. The great explosion referred to was engineered by General Pasley to blow up the Rounddown Cliff near Dover to make a roadway for the South-Eastern Railway rather than a more expensive tunnel; Sir John Herschel and several scientific men were to be present at this eagerly awaited spectacle.[18] O'Shaughnessy went on to say, 'It will afford me the utmost pleasure should I have the good fortune of meeting you on Thursday', which was the day of the planned explosion. There is no hard evidence as to whether he attended this event and met Sir John but reports in the press of the presence of Sir John and 'several scientific men' suggests that he may well have been there and met Herschel. Moreover, it is significant that soon thereafter, on 6 February, ten days after the explosion, he wrote to Herschel saying that at last he ventured on the risk of an election at the Royal Society, this suggesting that perhaps he did meet Sir John at the cliff demolition, who encouraged the young man in his ambition. The fact that General Pasley called on O'Shaughnessy is a mark of the latter's growing reputation but, of course, their shared expertise in engineering underwater explosions was undoubtedly a factor.[19]

General Sir Charles William Pasley (1780–1861) was a Scotsman, whose army career was chiefly concerned with engineering works but who achieved considerable fame in 1838 when he succeeding in producing underwater explosions which demolished sunken vessels which were an obstruction to shipping in the Thames; during six successive summers (1839 to 1844) he cleared the wreck of the *Royal George* from the anchorage at Spithead, and

[17] *Bombay Gazette*, 10 January 1840.

[18] 188 *Morning Herald*, 27 January 1843, p. 6, col. 1. 'The Grand Blast at the Dover Railway, Splendid Engineering Triumph' was the headline.

[19] RS, Correspondence of John Frederick William Herschel, letter from W.B. O'Shaughnessy.

96   AN INNOVATIVE PHYSICIAN AND SCIENTIST

that of the *Edgar* from St Helens on the Isle of Wight.[20] O'Shaughnessy's similar achievements in blowing up sunken vessels in the Hooghly had clearly impressed the general whose experience with explosives far exceeded that of the East India Company surgeon who had to improvise and did so very successfully. Whether O'Shaughnessy knew of Pasley's technique or whether typically he developed his own method is not known. It was a tribute to O'Shaughnessy's skill that Pasley invited him to view the removal of the cliff which was impeding progress on the railway extension to Dover, an honour for a man who was not military. The underwater explosions engineered by O'Shaughnessy to remove dangerous underwater wrecks on the Hooghly were obviously known to General Pasley, impressed him as an engineer and persuaded him to sponsor him for the Royal Society as a person who knew the candidate from general knowledge as was the case with Sir John Herschel.

First among the names of sponsors who had personal knowledge of O'Shaughnessy was that of Horace Hayman Wilson (1786–1860), assistant surgeon in the East India Company from 1808, assay master to the Calcutta mint and latterly professor of Sanskrit at Oxford University from 1832, leaving India to take up his duties in 1833. It is unlikely that Wilson and O'Shaughnessy crossed paths in Calcutta, as the latter arrived there only in December 1833, but their correspondence suggests that they had become friends, whether initially by exchange of letters or later in London when O'Shaughnessy was on medical furlough. Certainly, they met in London at the Royal Asiatic Society in 1843, where Wilson was president, possibly for the first time. The phrase 'promoter of education among the natives of Bengal' is interesting and may well explain the friendship which clearly had developed between O'Shaughnessy and Wilson. Although both had studied medicine, and both had been employed as assay masters at the Calcutta mint, the phrase is telling indicating that they shared an interest in education through the means of native languages, a major interest of Wilson and the reason why as a linguist he supported the Orientalist cause in the debates with the Anglicist faction in the late 1820s and early 1830s. These mutual interests apart, Wilson's commitment to Sanskrit and O'Shaughnessy's apparent neutral stance during the debate, together with his burgeoning interest in native medicine, would have brought them together, but there may also have been a reluctance on the part of O'Shaughnessy to support the supplanting of native languages entirely by English in education, as had happened in his native Ireland. Moreover, O'Shaughnessy was no mean linguist himself speaking Irish Gaelic, competent French and enough German to read scientific literature.[21]There is no question as to Wilson's dislike of the Anglicist cause and his support for the Orientalist position as typified by John

---

[20] R.H. Vetch, revised by John Sweetman, 'Sir Charles William Pasley (1780–1861)', *Oxford Dictionary of National Biography.*

[21] O'Shaughnessy was elected foreign secretary of the Medical and Physical

MEDICAL FURLOUGH IN LONDON AND THE ROYAL SOCIETY

Tytler (1790–1837), a Scottish surgeon in the East India Company service who was prominent in the ultimately unsuccessful movement to maintain and encourage education in Indian languages. O'Shaughnessy as far as can be ascertained was never vocal in his support for either cause, perhaps wisely as a very new arrival in Bengal hoping to further his career without antagonising either side. Despite the twenty-year difference in ages, the tone of their correspondence indicates friendship, particularly on reading O'Shaughnessy's relatively informal mode of address for Victorian times:

> My Dear Wilson,
>
> We are off tomorrow, and I write now to wish you farewell. My ballot at the Royal Society "comes off" on the 16 th. March. Can you endeavour to be present and make it a "dies albo notata (or nolanda) lapillo". Sir Charles is so much improved that I do not think it all likely that I may be detained in Canada later than July. John Grant's return home is… chiefly influenced by his hope of succeeding [illegible]. I hope he may for he is a really good fellow, a prince of masons and a kind friend. Don't forget the 16th – it would be irremediable disgrace to be rejected at the RS.[22]

Undoubtedly the two men had formed a close friendship, possibly through both being freemasons although there is no evidence available to suggest that Wilson was a freemason either in Bengal or Oxford. However, it is the last phrase in O'Shaughnessy's citation for election to the Royal Society, 'a promotor of education amongst the natives of Bengal', that may go some way to explaining the friendship despite Wilson being twenty-five years older. A further connection was that both had studied medicine, and both had been employed in assay at the Calcutta mint. The reference to John Grant as a 'a prince of masons' is evidence of a masonic connection and there is proof of this in the roll of a masonic lodge in Calcutta which O'Shaughnessy joined in 1844 soon after his return from furlough.[23] In their exchanges it appears as if Wilson was advising O'Shaughnessy as to the latter's future employment

---

Society on 2 January 1836 when he told the society that his only qualification as 'foreign' was that he spoke Gaelic.

[22] BL, H.H. Wilson Collection, MSS.EUR. E. 301/7, Letter from O'Shaughnessy to Wilson, 1 March 1843; John Peter Grant, later Sir (1807–1893) was an administrator in Bengal whose time in both Calcutta and in London between 1841-44 coincided with that of O'Shaughnessy. For an account of his life, see Kathleen Prior, 'Grant, Sir John Peter (1807–1893)', *Oxford Dictionary of National Biography* (2004). In the letter to Wilson, O'Shaughnessy referred to Grant as 'a prince among masons'.

[23] Roll of St John's Masonic Lodge Calcutta, William Brooke O'Shaughnessy initiated on May 23, 1844. I am grateful to Dr Margaret Makepeace of the British Library for this reference.

98 AN INNOVATIVE PHYSICIAN AND SCIENTIST

prospects when he returned to Bengal. Evidently, O'Shaughnessy did not view the prospect of his return to Bengal of being a mere regimental assistant surgeon with any enthusiasm.

O'Shaughnessy's letter of 1 March 1843 expressed the hope that Wilson would be present at the Royal Society ballot to ensure success, making it a 'dies albo notata (or nolanda) lapillo', literally a day marked by a white stone, a special day. Wilson had become a friend through their joint involvement in learned societies such as the Royal Asiatic Society, the Asiatic Society of Bengal and the Medical and Physical Society of Calcutta. Wilson was secretary of the Bengal Society from 1811 until he was appointed Professor of Sanskrit at Oxford in 1832 and later was director of the Royal Asiatic Society.[24]

The second name on the list of Fellows supporting his election 'through personal knowledge' is that of Colonel W.H. Sykes (1790–1872), soldier, politician, ornithologist, Indologist and one of the pioneers of the Victorian statistical movement, who, as a founder of the Royal Statistical Society, conducted statistical surveys examining the efficiency of army operations. He became a director of the East India Company in 1840 on his retiral from Indian Army service, was a member of the Royal Asiatic Society, president in 1858, elected FRS in 1834 and a council member on several occasions. Whether his personal knowledge of O'Shaughnessy's achievements arose from meeting him in India, or at the Royal Asiatic Society, is hard to say but as a member of the Court of Directors of the East India Company he would have been aware of O'Shaughnessy as a young scientist of note; his support as a Fellow of the Royal Society was important.[25]

Among his other sponsors are a group of distinguished medical men including two current physicians to Queen Victoria: Sir John Forbes and Sir James Clark. Sir John Forbes FRCP, FRS (1787–1861) was a distinguished Scottish physician and journalist, famous for his translation of the classic French medical text *De Auscultation Mediate (*1819) by René Laennec, the inventor of the stethoscope. His advocacy of the stethoscope, a new medical tool in the early years of the nineteenth century, was important, although the instrument was viewed at times with suspicion; its use in the diagnosis of diseases of the lungs, especially in consumption, was advocated by Forbes very early on and his support was undoubtedly influential in its eventual rather slow adoption in Britain. He was physician to Queen Victoria from 1841 until his death in 1861. O'Shaughnessy's ground-breaking work on the pathophysiology

---

[24] P. Courtright, 'Wilson, Horace Hayman (1786–1860), Sanskritist', *Oxford Dictionary of National Biography*, com.ezproxy.is.ed.ac.uk/view/10.1093/ref:odnb/9780198614128.001.0001/odnb-9780198614128-e-29657 (accessed 4 November 2023).

[25] B.B. Woodward, revised by M.G.M. Jones, 'Sykes, William Henry (1790–1872)', *Oxford Dictionary of National Biography* (2004).

of cholera during the 1831–1832 epidemic would certainly have been known to him, not least through his interest in medical journalism as joint owner and editor of the *British and Foreign Medical Review*, as a Fellow of the Royal Society and a willingness to embrace new medical advances. Another distinguished supporter was Forbes' friend, colleague and fellow Aberdonian, Sir James Clark, Bart (1788–1870), MD Edinburgh 1817, FRS 1832, who later became physician to Queen Victoria, reputedly an appointment that caused the Royal College of Physicians some disquiet, a reaction perhaps not dissimilar to that of the College when O'Shaughnessy first came to London. His care of John Keats, the poet, dying of tuberculosis in Rome has come in for criticism, some of which was perhaps largely unfounded.[26]

John Forbes Royle (1798–1858), surgeon and naturalist, was born in Cawnpore, India, the son of an army officer in the service of the East India Company. He followed his father in the service of the East India Company becoming an assistant surgeon in the Company's Bengal army in 1819. However, in common with many of the army surgeons, his main scientific interest was botany, eventually becoming superintendent of the botanical garden at Saharanpur where, unable to leave his post as an officer for field exploration, he employed plant collectors, 'and brought together a valuable collection of economic plants'. He became interested in the drugs sold in the local bazaars, his research forming the subject of his essay on the antiquity of Hindoo medicine, including an introductory lecture to the course of *materia medica* and therapeutics, later delivered as a lecture at King's College, London in 1837. In this text, Royle stressed the great antiquity and originality of Hindu medical texts and the fact that many ancient Indian remedies were worthwhile. It is not surprising that O'Shaughnessy in his preface to the *Bengal Dispensatory* refers to Royle as one of the authors from whose work the *Dispensatory* was chiefly compiled and undoubtedly his interest in and research into indigenous remedies would have been of major importance in the preparation of the volume. In 1836, Royle was appointed professor of *materia medica* at Kings College, London, becoming a Fellow of the Royal Society in 1837, eventually serving on the council. His botanical interests and his research into native medicines must have influenced O'Shaughnessy and in due course the latter's work on the Bengal dispensatory in turn persuaded Royle that O'Shaughnessy was a worthy candidate for the Royal Society Fellowship. It is also true that Royle's botanical research had persuaded the East India Company to explore the possibility of using local plants as the source of cheaper effective drugs, rather than importing them from Europe.[27]

[26] R.A.L. Agnew, 'Clark, Sir James, first baronet (1788–1870)', *Oxford Dictionary of National Biography* (2004).

[27] B.B. Woodward, revised by Mark Harrison, 'Royle, John Forbes', *Oxford Dictionary of National Biography* (2010); W.B. O'Shaughnessy, *The Bengal*

Dr Jonathan Pereira (1804–1853) is one of the most interesting of O'Shaughnessy's sponsors and perhaps also the closest to him in terms of their shared interests: chemistry, *materia medica* and pharmacology, where their impressive outputs of published papers matched each other, and moreover, both had published pharmacopoeia. In 1824, Pereira produced an English translation of the *Pharmacopoeia Londensis*, later becoming professor of *materia medica* in the new medical school in Aldersgate Street and lecturer in chemistry at the London Hospital. The *Medical Gazette* published his lectures on *materia medica* between 1835 and 1837, lectures which significantly were republished in India and these along with his *Pharmacopoeia* would have interested O'Shaughnessy.[28] His two-volume *Elements of Materia Medica and Therapeutics* is dedicated to Professor John Lindley, Professor of Botany in University College, London. As was mentioned earlier, O'Shaughnessy and Pereira had discussed in 1839 the potential of cannabis when he had been sent a cannabis plant by O'Shaughnessy, and later he referred to O'Shaughnessy's paper on cannabis in his *Elements*.[29]

The support of Sir James McGrigor, first Baronet (1771–1858), Director-General of the Army Medical Department from 1815, was important. He may not have been personally acquainted with O'Shaughnessy but there is absolutely no doubt that he would have known him by reputation. During the cholera epidemic of 1831–1832, Sir James was one of the men appointed to the Board of Health by the Privy Council in June 1831. As a Board member, he would have become aware of O'Shaughnessy early in 1832 when the young Irish doctor, not yet twenty-four, published the results of his chemical analysis of the blood and excreta of cholera victims from the north of England, writing to the Board of Health with his conclusions and recommendations for treatment. The reaction from the medical profession to the papers published in *The Lancet* and the later publication of Dr Latta's use of intravenous saline would have been sufficient to alert Sir James to O'Shaughnessy's original research and ten years later to convince him to lend his support. That a young man with original research to his name was unable to practise as a physician in London by the dictates of a biased cabal and was forced to leave London for India would have been known to Sir James; his successes in India would

---

Dispensatory and Companion to the Pharmacopoeia, published by order of the Bengal Government (London, 1842), in which Royle's works are mentioned.

[28] M.P. Earles, 'Pereira, Jonathan (1804–1853)', *Oxford Dictionary of National Biography* (2004).

[29] Jonathan Pereira, *Elements of Materia Medica and Therapeutics*, 2 vols (London, 1839–1840), vol. 2, dedication page to John Lindley; vol. 2, 1273 is a reference to O'Shaughnessy's work on cannabis.

have been impressive. A letter of support from a fellow surgeon and the senior officer in the Army Medical Department was an exceptional tribute.[30]

The final three names supporting his nomination may not have had the reputations of men like Horschel, Wilson or McGrigor, but all were men of distinction with notable achievements in various fields. Thomas Best Jervis (1796–1857) was a lieutenant-colonel of Irish extraction who served in the Bombay Engineers and had carried out surveys of the Indian subcontinent. O'Shaughnessy's experimental electric telegraphic line of 1839 must have impressed him and of course his published research on ancient Indian science and medicine was exactly the kind of work that would have brought him to the attention of O'Shaughnessy, but the Irish connection in this context may have had an importance as great than a joint interest in science and the value of indigenous science and medicine. The importance of family and national networks can never be underestimated: the role of the Irish in the development of the Empire has been examined by Crosbie, who all too accurately pointed out the scholarly neglect of this aspect of Irish imperial involvement of 'the reticence of scholars to examine the Irish within an Imperial context in India...'[31]

John Lindley FRS (1799–1865) was an English botanist and horticulturalist, perhaps the foremost of his time, elected to the Royal Society in 1828. O'Shaughnessy's awareness of the medicinal value of plants and his introduction of cannabis to Britain must have impressed Lindley. Lindley became professor of botany at London University in 1829.[32]

Thomas Horsfield (1773–1859) was born in colonial America, qualifying in medicine, where his doctoral thesis on the effects of poison ivy revealed his botanical leanings. He worked and travelled extensively in Indonesia, describing numerous species of plants from the region and later became a curator of the East India Company Museum in London, a post he retained until his death. He became FRS in 1828.[33]

The ballot for his election took place in London on 16 March 1843 at a time when O'Shaughnessy was en route to Canada as personal physician to Sir Charles Metcalfe. The Royal Society minutes record that O'Shaughnessy was one of four medical men to be approved as Fellows in the calendar year 1843, out of a total of twenty-two men elected. It is worth noting that among

---

[30] H.M. Chichester, revised by J.S.G. Blair, 'McGrigor, Sir James, first Baronet (1771–1858)', *Oxford Dictionary of National Biography* (2004).

[31] B. Crosbie, *Irish Imperial Networks. Migration, Social Communication and Exchange in Nineteenth Century India* (Cambridge, 2012), 3.

[32] R. Drayton, 'Lindley, John (1799–1865)', *Oxford Dictionary of National Biography* (2004).

[33] D.T. Moore, 'Horsfield, Thomas (1773–1859)', *Oxford Dictionary of National Biography* (2004).

the Council members at this time were two men prominent in the fields of photography and telegraphy, who would have been aware of O'Shaughnessy's experiments in the same fields: William Fox Talbot (1800–1877), photographic pioneer, and Charles Wheatstone (1802–1875), who developed the Wheatstone Bridge, an essential element in the development of telegraphy in England.[34]

It is remarkable and perhaps not coincidental that the title page of the *Bengal Dispensatory and Companion to the Pharmacopoeia*, published in London in 1842, records that among others the book is compiled from the works of Royle, Pereira and Lindley; significantly, all three were among O'Shaughnessy's sponsors for the Royal Society. Nevertheless, in this endeavour it was Horace Hayman Wilson who appears to have become his chief mentor and a frequent correspondent, later advising him as to his prospects when he returned to Calcutta after his sick leave. Despite his new distinction as a Fellow of the Royal Society, in India he remained as a mere assistant surgeon in the service of the East India Company. It may well have been Wilson's influence and support that secured him his renewed appointment to the post of chemical examiner to the government and that of deputy assay master to the mint on his return to India. He held these posts until 1851 when his involvement with telegraphy became of greater importance and will be explored in the next chapter.

In January 1844, O'Shaughnessy now an FRS, travelled with his wife and family, returning to India via Suez, the so-called overland route, completing the journey to Bombay on a paddle steamer, the Peninsular & Oriental Line *Hindostan*, in a considerably shorter time than the four-month voyage from London to Calcutta via the Cape of Good Hope on his first trip to India a decade previously. Technology was certainly changing the world of travel and before long O'Shaughnessy would be the man tasked with changing communications in India for ever, with consequences for British control of an increasingly discontented country.

---

[34] Records of the Royal Society. I am grateful to Rupert Baker, library manager at the Royal Society for the details of the number of fellows elected in 1843.

5

'That Man O'Shaughnessy' and Electric Telegraphy

The Times informed its readers in August 1852 of the proposal to develop an electric telegraph system in India to be supervised by Dr W.B. O'Shaughnessy:

> The East India Company have just determined to establish a very extensive system of electric telegraphs in India, under the superintendence of Dr W.B. O'Shaughnessy, of their medical establishment. It is intended to connect Calcutta, Agra, Lahore, Bombay, and Madras, and as many of the principal towns and stations as can be embraced in the routes between these places. The distance to be traversed is upwards of 3,000 miles, and it is intended to proceed with such expedition in its construction that its completion may be expected before three years from the present time. Dr. O'Shaughnessy has lately been employed in India in carrying on experiments with the electric telegraph, in order to discover the best system which could be adopted. The result of these experiments was highly satisfactory to the Governor-General and to the Court of Directors, who immediately resolved to take measures for giving to India the inestimable advantage of this marvellous means of communication.[1]

The purpose of this chapter is to examine the history of Indian telegraphy, as established and supervised by O'Shaughnessy: first, his work will be placed in the context of the advance of the electric telegraph in Europe and America; secondly, it will focus on the part played by Dalhousie in promoting and establishing the telegraph network in India; and finally, the contribution of O'Shaughnessy in constructing this network will be examined, analysing his role from the time of his early Calcutta experiments in 1835 to his appointment in 1852 as superintendent of electric telegraphs. The final section will examine the question as to what extent the telegraph lines helped the British army to deploy troops and defeat the rebels during the insurrection of 1857, a question that has divided observers since the time of the uprising itself.[2]

---

[1] *The Times*, 17 August 1852, p. 5, col. 1.

[2] The term commonly used to describe the rebellion of 1857–1858 has been mutiny. This is pejorative, suggesting that it was purely a military insurrection, perhaps a perception that suited British Imperial propaganda during its occupation of India but

At a Calcutta reception in honour of Lord Dalhousie on his departure from India at the conclusion of his term as governor-general the following sentiments were expressed by the Honourable James Hume, a Calcutta magistrate, journalist and fellow Scot:

> As long ago as Lord Auckland's time the distinguished officer who has done more for electrical communication in India than has ever been accomplished in a like period in any part of the world, demonstrated its feasibility on a wire of some five miles in length, but shall we less grateful to the man, under whose rule and by whose determined will, this wonderful means of communication has been so rapidly carried out.[3]

This linking of Lord Dalhousie and O'Shaughnessy, drawing attention to their joint development of electric telegraphy in the context of a farewell dinner to the governor-general was understandable, but the fulsome praise for O'Shaughnessy was perhaps unusual on such an occasion. Of course, Dr William Brooke O'Shaughnessy is not mentioned by name but of the 500 dignitaries present in Calcutta Town Hall on that august day in March 1856, few would have been unaware of the man who was the recipient of such praise. That the two men were linked is not surprising, for without the enthusiasm of Dalhousie for the project and the chance that such a man as O'Shaughnessy with experience in telegraphy happened to be in Calcutta at the right time, it is doubtful if anything useful could have been achieved, although army engineers might have other views.

A century or so later, an independent India would celebrate a hundred years of electric telegraphy with an official centenary book which included the following words:

> We would do well to remember quite how revolutionary the telegraph was when it emerged as a major tool of communication in the mid-nineteenth century. Regions, countries, and empires could all be brought together virtually with a series of dots and dashes that translated into a means of informing, and of course, *controlling societies.* [author's italics][4]

---

avoids the reality that the civilian population were also involved. In this context, the word mutiny will not be used.

3 NRS, Dalhousie Papers GD 45/6/218, *Meeting in Honour of Lord Dalhousie in Calcutta Town Hall* (Calcutta, 1856), 24, 25. The name Hume raises the question as to whether he was a kinsman of the Joseph Hume who chaired the London College of Medicine pressure group; It is unlikely that O'Shaughnessy was present at this celebration – in late 1855 he was given permission to revisit Europe and America to assess the Morse system and to recruit signallers who were expert in the use of Morse code.

4 R. Mitter and A. Iriye, 'Foreword', in D.K.L. Choudhury, *Telegraphic Imperialism. Crisis and Panic in the Indian Empire c. 1830* (Basingstoke, 2010), ix.

'THAT MAN O'SHAUGHNESSY' AND ELECTRIC TELEGRAPHY

This anniversary publication was published by the Posts and Telegraph Department of the Indian Government in 1953, with a foreword by Jawaharlal Nehru (1889–1965), the first prime minister of an independent India and an ardent anti-colonial nationalist. Nehru avoided any mention of the British occupation in his foreword, perhaps preferring (understandably) to talk in general terms saying:

> The telegraph, even more than the railway, brought the new method of rapid communication. (Probably the most remarkable feature of the age we live in is the rapidity of communication). Starting a little over a hundred years ago with the telegraph, it has spread in many ways – the telephone, the wireless and, lately, radar. Nothing has changed the world more during these hundred years than this astonishing change in our methods of communication. The telegraph was the first great step in this direction. Let us, therefore, honour the telegraph as the herald of the New Age.

Krishnalal Shridharani, the author of the centenary volume, wrote of how 'thousands of miles apart the first experimental telegraph lines were constructed in India and America, the oldest and the most modern, in the same year, 1839. The pioneer in India was Sir William O'Shaughnessy Brooke, fondly remembered as Dr O'Shaughnessy.' The author headed chapter five of his monograph with the words 'That Man O'Shaughnessy', a phrase used in this chapter heading also. It is noteworthy that one hundred years on and seventy years after O'Shaughnessy's death, his memory was fondly remembered by telegraphists despite rebellions and the subsequent battles for independence from Britain.[5] Shridharani records how O'Shaughnessy accidentally discovered 'by the falling of a wire into a large tank at the Medical College', that when water was available only one insulated wire was required to transmit; whether this account was true is questionable. Further trials were carried out, centred on the Botanic Gardens with wire immersed in the Hooghly and ending eventually in Dr Wallich's library in his house in the gardens. Excellent transmission was recorded. A further trial using over 3,000 yards of wire was carried out on this occasion in Sir John Royd's garden, again with perfect transmission. These experiments became the foundation of India's extensive electric telegraph system.[6] Shridharani was eager to highlight

---

5   Krishnalal Shridharani, *Story of the Indian Telegraphs. A Century of Progress* (New Delhi, 1954), 2; the use of Brooke as the surname is accurate because in 1861 he changed this name by royal assent to O'Shaughnessy Brooke, for reasons which will be explained in a later chapter.

6   Dr Nathaniel Wallich (1786–1854), a physician and botanist, was superintendent of the Calcutta Botanic Gardens, a friend and colleague of O'Shaughnessy and of Robert Graham, Professor of Botany in the University of Edinburgh, whose classes O'Shaughnessy attended. For more on his career, see Roger de Candolle and Alan Radcliffe-Smith, 'Nathaniel Wallich, MD, PhD, FRS, FLS, FRGS (1786–1854) and

the contribution of Indian expertise: 'And yet nationhood in its modern sense that we have achieved is in no small measure due to the beginnings made in Calcutta by one Dr O'Shaughnessy later to be supplemented by such Indian associates as Seebchunder Nundy...'[7] Seebchunder Nundy (1824–1903), who was born in poor circumstances in Calcutta, joined the refinery department of the Calcutta mint at the age of twenty-two in 1846 at a time when Dr O'Shaughnessy was the chemist and assistant assay master. His abilities were soon noticed, and he was selected to be O'Shaughnessy's personal assistant. When O'Shaughnessy became superintendent of telegraphs, he again chose Nundy to be his aide.

Dalhousie's delight and wonder in the new technology is unmistakeable. In a letter written in November 1854 to his old friend, Sir George Coupar (1788–1861), 1st Baronet, he told him of his amazement at the rapidity with which news now travelled: a telegraphic message arrived in Calcutta the previous evening, intimating that Lord Fitzclarence had died in Bombay that morning at 10 o'clock. He wrote, 'there was something awesome in this abrupt announcement of sudden death at a great distance on the very day in which it occurred'.[8] The instigator of the telegraph in India was himself awestruck by the rapidity of the new technology; later he would tell Parliament in a final minute recording events during his last months as governor general, of the efficiency and rapidity with which the system was put in place by Dr O'Shaughnessy. It was Dalhousie himself who described the railways, electric telegraph and postage as 'the three great engines of social improvement', all three he was determined to establish in India, and it was Dr William Brooke O'Shaughnessy to whom he turned for the delivery of telegraphy.[9]

It is important to stress that O'Shaughnessy was working in relative isolation from Western academic centres, developing his own techniques as early as 1835, but no doubt reading about developments in scientific journals. His skill, his ingenuity and perseverance remain worthy of admiration bearing in mind that it is generally accepted the electric telegraph was introduced in Britain more or less jointly by William Fothergill Cooke, later Sir William (1806–1879) and Charles Wheatstone, later Sir Charles (1802–1875), whose

---

the Herbarium of the Honourable East India Company, and their relation to the de Candolles of Geneva and the Great Prodromus', *Botanical Journal of the Linnean Society*, 83, 4 (December 1981), 325–348.

   [7] Shridharani, *Story of the Indian Telegraphs*, 2.

   [8] James Andrew Broun Ramsay Dalhousie and J.G.A. Baird, *Private Letters of the Marquess of Dalhousie* (Edinburgh and London, 1910).

   [9] 213213 Suresh Chandra Ghosh, *Dalhousie in India, 1848–56. A Study of his Social Policy as Governor General* (New Delhi, 1975), 2.

patent for a telegraph apparatus was granted in June 1837.[10] In July of that year, they went on to demonstrate the effectiveness of their system with a nineteen-mile experimental circuit between Euston and Campden Town on the London and Birmingham railway and in the following year 'the first actual working telegraph' was erected between Paddington and West Drayton on the Great Western Railway.[11] These early English developments were constructed along established rail tracks, deliberately so as an aid to signalling in the prevention of accidents, particularly on single track lines where accidents and demands for increased safety were fast becoming an issue. The contrast between the relatively straightforward process of constructing telegraph lines alongside an established track in an urban setting and the difficulties faced by O'Shaughnessy in Bengal, working in untamed territory that had never been surveyed, is striking and will be discussed later.

The work of Cooke and Wheatstone was known to O'Shaughnessy despite being remote from Europe and America, but of course he was well aware of their early work in electromagnetism: in a series of lectures on natural philosophy, later recorded in a memoir, he described the method used by Cooke and Wheatstone in which:

> Five dipping needles are employed, which requires six wires to work them, and which by combined movements of two or more needles give every variety of signal which can be required. The wires are covered with an insulating material and are all placed for security in an iron tube led above the ground from station to station.

The cost of each wire, O'Shaughnessy wrote, was £7 per mile, but when the cost of the iron tube protecting the wires was added, the cost rose to between 250 and 300 pounds per mile. He went on to say that 'nothing could be more perfect in its action than this telegraph' but somewhat inconsistently at the same time he was critical of the multiplicity of wires, arguing that two wires at most would do and where a railway or canal existed one wire would be sufficient. It is difficult to reconcile both views.[12] He was never afraid to criticise constructively what he considered to be faulty or inferior.

---

[10] James Burnley, revised by Brian Bowers, 'Cooke, Sir William Fothergill (1806–1879)', *Oxford Dictionary of National Biography* (2004); S.P. Thomson, revised by Brian Bowers, 'Wheatstone, Sir Charles (1802–1875)', *Oxford Dictionary of National Biography* (2004).

[11] R.S. Culley, *A Handbook of Practical Telegraphy* (London, 1874), v.11. A frontispiece reads 'Adopted by the Post Office and by the Department of Telegraphy for India'.

[12] M. Adams, *Memoir of Surgeon- Major Sir William O'Shaughnessy Brooke in Connection with the Early History of the Telegraph in India, Compiled by Permission*

108 AN INNOVATIVE PHYSICIAN AND SCIENTIST

The experimental line in London was not the only one being tried out in Britain, for at roughly the same time as Cooke and Wheatstone were building their line, another one, little known then or now outside of Scotland, was being demonstrated in Edinburgh by William Alexander in the summer of 1837 with a four-mile experimental line in the grounds of the university. As far as can be ascertained, it was only reported in the Scottish press which described this experiment in telegraphy, writing in early July 1837 and at the same time referencing their recent article on an 'ingenious invention' and the work of Wheatstone:

> Electric Telegraph. – When we mentioned an ingenious invention on Saturday and alluded to the experiments made by a scientific gentleman in London, we were not aware that Mr W. Alexander, of this city, with the aid of Mr Kemp, had been engaged for some weeks past in an elaborate series of experiments on the same subject, in Dr Hope's Class-room. We had the pleasure of witnessing there on Tuesday the transmission of the electric action through four miles of copper wire. We saw it act instantaneously on the magnetic needle, produce sparks, and explode a detonating powder, not merely at the termination, but at various parts of the circuit of four miles. The experiments have been repeated in the presence of Professors Hope, Jameson, and Traill, the Dean of Faculty, the Commander-in-Chief (Lord Greenock), the Lord Provost, and a number of other gentlemen. We shall advert to the subject again on Saturday. The scientific gentleman in London, whose telegraph we described in our last, is Professor Wheatstone of King's College.[13]

The Mr Kemp mentioned as assisting W. Alexander, was Kenneth Treasurer Kemp, who died in November 1842 at the age of thirty-seven while a lecturer in practical chemistry in the University of Edinburgh. His obituary notice included the following item:

> During the earlier part of his course as a teacher of chemistry, he devoted his attention chiefly to voltaic electricity, and produced many interesting and valuable modifications of galvanic apparatus; and to him the world is indebted for the advantages derived from the amalgamising of the zinc plates-an improvement entirely his own, and one which is now almost universally adopted in the construction of galvanic batteries.[14]

---

of the Director – General of Telegraphs (Simla, 1889), Appendix: Lecture Seventh, 22.

[13] *Fife Herald*, 6 July 1837, p. 3, col. 7; sadly, nothing is known or can be discovered about William Alexander; Mr Kemp was described as Experimental Assistant to Professor Hope, the professor of chemistry.

[14] 'Death of Mr. K. T. Kemp, Lecturer on Practical Chemistry in the University of Edinburgh', *Provincial Medical Journal and Retrospect of the Medical Sciences*, 5, 119 (1843), 302; J.B. Morell, 'Practical Chemistry in the University of Edinburgh

Kemp's original work on galvanic batteries was published at length in *The Edinburgh New Philosophical Journal* of 1828, and it is worth noting that Fahie in his important and detailed history of electric telegraphy credited Kemp of Edinburgh for his work on batteries. Kemp certainly was at the forefront of research into the possibilities of telegraphy but sadly his name has largely disappeared from the annals of electric telegraphy research.[15] His paper of 1828 was published in Edinburgh at a time when O'Shaughnessy was a medical student there with a deep interest and ability in chemistry and the expanding new science of galvanism. It is perfectly possible that Kemp's work triggered and developed O'Shaughnessy's interest in telegraphy; the two men were of similar ages, devoted to scientific experiment, both involved in chemistry research, teaching chemistry in Edinburgh, and it is inconceivable that they did not know each other. O'Shaughnessy as an interested contemporary was aware of Kemp's research and publications as the following extract reveals. In an address to the Medical and Physical Society of Calcutta in 1837, later published, on the topic of galvanic batteries, he said, 'the first important step towards the discovery of a remedy for these evils was accomplished by Mr Kemp of Edinburgh in 1829'. The evil referred to was the rapid destruction of the zinc plates 'as to render the battery useless'. Kemp found that an amalgam of mercury and zinc associated voltaically with copper acted better than zinc alone. In his opening remarks on this occasion, O'Shaughnessy acknowledged the financial help he had received to enable him to continue with his experiments on galvanic batteries, among those he thanked were the governor-general, the Medical and Physical Society and leading members of the Society.[16] Such was the interest in practical chemistry in the late 1820s and early 1830s that no fewer than four lecturers were active in Edinburgh at this time, possibly one of the reasons why O'Shaughnessy made the decision to leave for London, the field being too crowded and as an outsider he may have felt excluded and unlikely to achieve success ahead of his Edinburgh peers.

1799–1843', *Ambix*; in a footnote on p. 77, Morell records that in 1835 there were four courses available in practical chemistry in Edinburgh, including one conducted by Kenneth Kemp. Comrie's *History of Scottish Medicine* includes William Gregory, D.B. Reid and Andrew Fyfe as lecturers in chemistry but strangely makes no mention of Kemp. As far as can be discovered Kemp did not study or qualify in medicine.

[15] 'Description of a New kind of Galvanic Pile, and also of another Galvanic apparatus in the form of a Trough. By Mr Kemp. With Plates. Communicated by the Author', *The Edinburgh New Philosophical Journal*, 6 (October 1828–March 1829), 340–348; John Joseph Fahie, *A History of Wireless Telegraphy* (Edinburgh and London, 1901).

[16] W.B. O'Shaughnessy, 'Experimental Enquiries on the Laws, Practical Improvement and Useful Applications of the Galvanic Battery', *Quarterly Journal of the Calcutta Medical and Physical Society* (1 October 1837), 484–507, 484, 487.

110   AN INNOVATIVE PHYSICIAN AND SCIENTIST

The *Caledonian Mercury* was another Scottish newspaper which took up the theme, developing the topic of electric telegraphy, on this occasion referring to events in Germany:

> In the annals of invention, it is generally found that new ideas, arising out of the progress of science, occur to different individuals simultaneously. While our towns-man, Mr Alexander, and Professor Wheatstone of London, were applying the powers of Galvanism to the purpose of internal communication, we find that a German professor was engaged with similar experiments. These wires are intended to exemplify a project of Professor Steinhill, for the conveyance of intelligence by means of electric magnetism. it is stated, that, in two seconds communication might conveyed from Lisbon to St Petersburg, by means of a telegraph of this description.[17]

Jeffrey Kiev in his splendid history of the electric telegraph refers to Alexander as a Scottish inventor who shared his ideas in *The Times*, showing the practicability of his scheme for telegraphy – not only the science but ideas for its use and the estimated costs. Alexander later also opposed the application of Cooke and Wheatstone for a Scottish patent but eventually withdrew 'acknowledging the superiority of Cooke and Wheatstone's plans'. The wide interest in this new technological advance is clear when the press was highlighting experiments in Edinburgh, London and Germany. Alexander took his ideas further, proposing how this new means of communication could be used to greater general benefit and suggested his scheme could be adopted all over the country, going so far as to write to the Treasury advocating government patronage, patronage that was to be forthcoming in India under Dalhousie but did not happen quite as soon in Britain.[18]

William Alexander seems to have been the main promoter of telegraphy rather than Kemp, for in June 1837 he wrote to Lord John Russell, the home secretary, suggesting the construction of an electromagnetic telegraph line between Edinburgh and London. His proposal of an Edinburgh–London link he also communicated to the press: in a letter to the *Scotsman* newspaper headed 'For an Instantaneous Telegraphic Communication betwixt Edinburgh and London by means of Electric or Voltaic Currents Transmitted through Electric Currents underground', he related the success of the four-mile experimental line in the university grounds, listing the scientists and dignitaries who had viewed the occasion and detailed a plan 'to communicate [the plan] to His Majesty's government as worthy of their attention, in place of being left to the joint enterprise of individuals...' Alexander suggested that the telegraph

---

[17] *Caledonian Mercury*, 22 July 1837, p. 1, col. 3.

[18] Jeffrey Kieve, *The Electric Telegraph. A Social and Economic History* (Newton Abbot, 1973), 24.

wires should run beside the trunk road from Edinburgh to London, buried to a depth of two to three feet; clearly, this was not a scheme designed purely for railway signalling and safety but intended for political and commercial benefit, more akin to that which was later developed in India as a means of communication, not merely for the benefit of the railways.[19]

It is worth noting that although steamboats and railways in India were mostly developed through private enterprise, the telegraph in India was entirely financed by government, a system similar to that recommended by Alexander in his letter to *The Scotsman*. It could be construed that this public letter was an attempt by Alexander to obstruct or delay the presumably private enterprise schemes of Cooke and Wheatstone, perhaps to promote his own scheme or was he being altruistic hoping to encourage official action? His attempt to involve the government was in the end totally unsuccessful, with no trace of any subsequent government interest or action. This is in distinct contrast to the sponsored development of Indian telegraphy financed by the East India Company, essentially a British government offshoot, and to the apparent lack of action at home when, despite these early developments and protestations, ten years elapsed before Parliament authorised the formation of the first telegraph company on 18 June 1846.[20] Of course, it happened in India for very specific reasons, reasons political, military and economic, that did not apply in Britain at the time; Ghose in an excellent appraisal of telegraphy in India makes the point that 'political and military necessities outweighed social and economic considerations in the development of the electric telegraph in India', an aspect that will be explored later.[21]

Across the Atlantic in America, two men, Joseph Henry and Samuel Morse, were at the forefront of developments in electromagnetism. Joseph Henry (1797–1878), the son of poor Scottish immigrants, developed the electromagnet and the concept of the electric relay into what in time became a practical device. Although Henry himself never built a functioning telegraph line, his work was crucial in laying down the scientific basis for future practical developments; he was working alone on his research during the early 1830s at the same period as Cooke and Wheatstone in London and Kemp in Edinburgh but later elaborated his ideas considerably further.[22] According to O'Shaughnessy,

---

[19] *The Scotsman*, 1 July 1837, p. 2, cols 4, 5. The timing of these three reports and the dates of the trials are inconsistent.

[20] Simone Fari, *Victorian Telegraphy before Nationalisation* (London, 2015), 12; the East India Company was nationalised in 1858.

[21] Saroj Ghose, 'Commercial Needs and Military Necessities: The Telegraph in India', in *Technology and the Raj: Western Technology and Technical Transfers to India 1700–1947* (New Delhi, 1995), 153, Ghose's doctoral thesis was on the topic of the electric telegraph in India.

[22] Thomas Coulson, *Joseph Henry: His Life and Work* (Princeton, NJ, 1950), 9;

Henry in 1838 proposed to employ the sudden development of magnetism in a horseshoe-shaped bar of soft iron, attracting a light piece of iron carrying an arm which when attracted marked dots on a revolving cylinder. O'Shaughnessy was again critical, on this occasion of the method used by Henry where eleven miles of wire were coiled *spirally* round a cylinder, a system which in O'Shaughnessy's opinion invalidated the results.[23]

Samuel Finley Breese Morse (1791–1872), the son of a Calvinist preacher, was a successful painter who studied in England between 1811 and 1815 and who had a very successful career as a portraitist. As the result of a chance meeting with a knowledgeable telegraph enthusiast on a transatlantic voyage back to America, he became fascinated by the science of electricity and telegraphy. He soon developed the concept of the single wire telegraph, first demonstrating its capability in January 1838. His claims to have been the sole inventor of the electric telegraph are highly controversial, sparking off a bitter feud with Joseph Henry when Morse in a publication disparaged Henry's right to be acknowledged as one of the scientists whose work was the basis of a functioning telegraph. An enquiry in 1858 set up at Henry's behest by the Smithsonian Institute came down unequivocally in his favour, being highly critical of Morse.[24] However, in 1858, before this quarrel erupted the Morse telegraphic apparatus had been adopted for European telegraphy and of course his eponymous code eventually became the standard worldwide. The name of Henry is now little known except among a few telegraphy enthusiasts.

Seventeen years had elapsed from the time O'Shaughnessy as a twenty-six-year-old, recently arrived in India, began experimenting in 1835 with galvanism and electricity to his appointment as superintendent of electric telegraphs in India in 1852. As *The Times* reported, Dr W.B. O'Shaughnessy, only three years after his appointment to the Bengal medical service, had begun to experiment in 1837 with electric telegraphy, although earlier he had conducted experiments on galvanic batteries and electricity soon after his arrival in Calcutta. As early as July 1835, at a so-called 'Scientific Party' in Government House, he gave a lecture and demonstration on the properties of galvanic electricity

---

Extracts from the proceedings of the Board of Regents of the Smithsonian Institution, in relations to the electro-magnetic telegraph (Washington, DC, 1858). The Board were asked by Henry to examine statements made in an article by Morse in *Shaffner's Telegraph Companion* which Henry considered to be erroneous and defamatory; the Board believed and said so that the piece was little more than an assault on Henry, attempting to deprive him of his honours as a scientific discoverer and attacking his integrity.

[23] Adams, *Memoir of Surgeon-Major Sir William O'Shaughnessy Brooke*, Appendix, p. 23.

[24] *Extract from the proceedings of the Smithsonian Institute, Letter from Joseph Henry to the Board of Regents*, 5–8.

and demonstrated the power of his galvanic battery which he claimed, as a result of improvements he had personally effected, had greater power than the most efficient batteries in Europe.[25] His Scientific Party was not dissimilar to the spectacles enjoyed by the middle and upper classes in London at this time. Morus writes, 'There were senses in which the telegraph did not require invention. Its components were already common features of any electrician's laboratory: the electromagnets, the galvanometer needles, the voltaic batteries that made up an electrician's working equipment.' For many, early Victorian electricity was a matter of display. William Sturgeon and his cohorts at the London Electrical Society, for example, devised a range of devices for making electricity spectacularly visible to their audiences. O'Shaughnessy was a member of the London Electrical Society, and he had become adept at displaying the possibilities of electricity, both in entertainment value and as a practical undertaking.[26]

In 1841, O'Shaughnessy wrote a detailed paper using extracts from his notes of lectures 'On Galvanic Electricity' and 'On the Charcoal Light', all of these lectures reproduced by Adams in his memoir of 1889. He recounted his first attempt to prove that a long line transmitting electric impulses was feasible and in 1837 he constructed in the Calcutta Botanic Gardens a telegraph line of iron wire about 2 mm thick supported by 42 rows of bamboo poles, the lines passing backwards and forwards, making a complete circuit of 30 miles. This was the first telegraph line constructed in the East, remarkably in the same year that similar trials were being conducted in Edinburgh and London. Over this line he carried out a series of experiments to establish a simple yet practical method of receiving the electric signal.[27]

In the definitive published description of this remarkable experiment, he described first how one terminal was placed in the house of Dr Nathaniel Wallich, the gardens' superintendent, whose help O'Shaughnessy acknowledged and then he was able 'to construct a line of wires of sufficient length to afford practically valuable results'; how with Dr Wallich's 'liberal aid' a

---

[25] *The East India and Colonial Magazine*, 11 (1835), 8–710, 569–570.

[26] Iwan Rhys Morus, 'The Electric Aerial: Telegraphy and Commercial Culture in Early Victorian England', *Victorian Studies*, 39, 3 (1996), 339–378, 342; see also Iwan Rhys, Morus, *Frankenstein's Children: Electricity, Exhibition, and Experiment in Early-Nineteenth-Century London* (Princeton, NJ, 2014). In this context, electrician refers to a person with an academic interest in electricity and its potential. The Royal Asiatic Society of London records in its minutes for the 1830s and 1840s regular presentations to the Society of publications from the London Electrical Society, items which sadly are not preserved.

[27] J.A. Bridge, 'Sir William Brooke O'Shaughnessy, M.D., F.R.S., F.R.C.S., F.S.A.', *Notes and Records: The Royal Society Journal of the History of Science*, 52 (1998), 109.

parallelogram of ground was planted with forty-two lines of bamboos, each fifteen feet in height. Each row was arranged so as to have half a mile of cable in one continuous line, the wires employed being of iron, one twelfth of an inch in diameter. He explained that iron was used because of cheapness but the results he thought would still be of value; iron was one of the poorest of the metallic conductors of electricity and therefore it seemed reasonable that if it worked with iron it would work with copper, one of the best conductors. The total circuit was twenty-two miles.

Several experiments with transmission were carried out including what was described as 'Mode of Correspondence by Pulsations and Chronometers', the pulsations being simply electric shocks experienced by a person holding metallic handles in which the wires ended, and the chronometers functioned as a means of recording the pulses in such a way as to be intelligible when the second hand was deflected in recognisable patterns. At seven miles distant, the shocks were 'exceedingly smart' but diminished as the distance increased. He then experimented with transmission by water, using eleven miles of metal and 13,256 feet of water circuit finding that the signals passed 'as intelligibly and strongly as before'. O'Shaughnessy, in his typical fashion, discussed in the manual in detail the history and principles of the electric telegraph writing that 'the source of electricity of most importance to the telegraphic student is that exerted by chemical action on metals. It is termed Voltaic or Galvanic from Volta and Galvani, two illustrious Italian philosophers.'

The first report of this experiment was published in the journal of the society of which O'Shaughnessy was officiating joint secretary, yet another indication of his versatility and engagement in academic affairs. He wrote, 'there are few projects which at first sight appear so visionary as those which promise practical benefits to mankind through the agency of electrical operations'. It was to be a decade before he was to be given the opportunity to carry out such a visionary project although undoubtedly the benefits were to be at the outset largely limited to the colonial ruling power both politically and commercially and, of course, later militarily. Moreover, he declared that the experiments seemed to him to be conclusive as to the practicability of establishing at a cheap rate 'telegraphical communications' which would be 'perceptible by night and day and in all varieties of weather and season'.[28]

Choudhury, in a wide-ranging paper, discussed the role of the Asiatic Society in some of the early attempts to promote electric telegraphy in India. For example, a highly ambitious proposal in 1839 for an electric-hydraulic telegraph was put before the Society by Adolphe Bazin, Baron du Chonay,

---

[28] W.B. O'Shaughnessy, 'Memoranda Relative to Experiments on the Communication of Telegraph Signals by Induced Electricity', *Journal of the Asiatic Journal of Bengal*, 8 (1839), 714.

'for effecting correspondence between Calcutta, London and the rest of the world'. The Society minutes record that M. Bazin, referring to his experiments, 'admitted freely that not one of the electrical machines... could be made to produce the least sign of excitement', hardly an encouraging recommendation.[29] O'Shaughnessy, in September 1839, presented a detailed monograph to the Society on the history and present status of electric telegraphy in the course of which he pointed out the faults and drawbacks in M. Bazin's scheme, foremost of which were the problems of humidity and moisture experienced in Bengal that a non-insulated scheme would experience:

> There are few projects which at first sight appear so visionary as those which promise practical benefit to mankind through the agency of electrical operations. From the dawning of knowledge in this science, pretenders of every grade have found it a free field for their speculations: and hence perhaps it arises that the sober and practical part of society generally regard with distrust, the multitudes of projects which electricians are constantly advancing. We nevertheless find that many eminent philosophers – whose habits of cautious research, have been proved by their numerous contributions to the mass of general science – such men as Brande, Faraday, Wheatstone, and Fox – are amongst the foremost, who predict many real advantages to the community from the application of the mysterious, though readily controllable forces which electricity places at our command.[30]

His account, entitled 'Memoranda relative to experiments on the communication of Telegraphic Signals by induced Electricity', extended to seventeen pages of well-argued concepts, concluding thus:

> They appear to me conclusive as to the perfect practicability of establishing, at a cheap rate, telegraphical communications, acting through electrical agencies, certain and infallible in their indications, perceptible alike by night and day, in all varieties of weather and season, and, lastly, so swift in their nature, that the greatest distances concerned bear scarcely any appreciable proportion to the inconceivably brief period in which the signal can be conveyed.[31]

---

[29] *Journal of the Asiatic Society of Bengal* (1840), 436–437, minutes of the meeting of the Society, May 1839; Deep Kanta Lahiri Choudhury, '"Beyond the Reach of Monkeys and Men?": O'Shaughnessy and the Telegraph in India c. 1836–56', *Indian Economic and Social History Review*, 37, 3 (2000), 259–381. Adolphe Bazin remains a shadowy figure in the history of electric telegraphy with biographical details unobtainable in English or French records.

[30] *Journal of the Asiatic Society of Bengal* (1840), 714–731, 731.

[31] O'Shaughnessy, 'Memoranda', *Journal of the Asiatic Society of Bengal* (1840), 714.

Choudhary quotes a letter from O'Shaughnessy written to J.P. Grant, secretary to the government of Bengal, in which he described himself as the man 'who in 1837 declared an electric telegraph to be a practical thing... [and] I proved it so in 1838'.[32]

An earlier chapter described how, on his return from furlough in England in 1844, O'Shaughnessy took up his old post of chemical examiner to the Bengal government, also acting as deputy assay master of the Calcutta mint from November 1844 to January 1851 when he was promoted to assay master. After fifteen years as an assistant surgeon, notwithstanding his remarkable elevation to the Fellowship of the Royal Society, it was only in 1848 that he was promoted to the rank of surgeon, length of service being the criterion rather than honours or distinctions.[33] He remained involved in chemical analysis, the position of chemical examiner requiring forensic and chemical expertise, both subjects he had studied in Edinburgh under distinguished exponents such as Robert Christison (1797–1882), later Sir Robert Christison, Bart, in forensic medicine, and Thomas Hope (1766–1844) in chemistry. He published some of the results of his investigation into cases of actual or suspected poisoning but by the end of the decade, with the appointment of a new governor-general, his scientific interests were to change. The arrival in July 1847 of James Ramsay (1812–1860), Lord Dalhousie, as governor-general of India heralded a major change in Indian affairs and one that had huge repercussions for O'Shaughnessy's career.

The appointment of Dalhousie, a Scottish earl, as governor-general of India in January 1848 in succession to Lord Hardinge of Lahore (1785–1856), who was sixty-three when he left India, saw a younger man take over, one who, although not in the best of health, saw the benefits of new technologies in terms of rapid communication whether by rail or telegraphy and was determined to follow these concepts through to fruition. Dalhousie succeeded to his Scottish earldom in 1838 and therefore had to leave the Commons for the Lords, a change which in no way hindered his progress in government, successively becoming vice president of the Board of Trade and then president, succeeding Gladstone in this role. Gorman wrote, 'it is certain that all of the expertise and enthusiasm of O'Shaughnessy would have been to no avail were it not for the advent of the regime as governor-general of India of James Andrew Ramsay, Marquis of Dalhousie'. Before leaving England, Gorman points out,

---

[32] Choudhary, 'Beyond the Reach of Monkeys and Men?', where he quotes from Home Department, Public Proceedings, 23 April 1852, No. 13: From W.B. O'Shaughnessy, Superintendent of the Electric Telegraph, to J.P. Grant, Secretary to the Government of Bengal, letter 10 February 1852, National Archives of India.

[33] Adams, *Memoir of Surgeon-Major Sir William O'Shaughnessy Brooke*, 5, 6.

Dalhousie had already gained useful experience of railways and telegraphy and 'was a progressive advocate of industrial and technological change'.[34]

Dalhousie left London on 11 November 1847, travelling via Alexandria, the overland route, to arrive in Madras on 5 January 1848, a journey considerably shorter than that taken by his protégé, O'Shaughnessy. His biographer, Sir William Lee-Warner, wrote that 'he bequeathed to those who came after him not only a reorganised system of administration, but a length of nearly 4000 thousand miles of telegraph connecting Calcutta with Peshawar, Madras and Bombay'.[35] But these were not his only achievements: he was the archetypal empire-builder and it will be argued that this urge led him to see that improved communications while expanding the Empire's borders were going to be crucial to political and military control. This drive to extend the frontiers of the Empire was made clear in a letter to a friend where he wrote: 'meantime I have got two other kingdoms on hand to dispose of – Oude and Hyderabad. Both are on the high road to be taken under our management...'[36] These sentiments appear now as astonishingly cavalier and carefree about the fate of several million fellow human beings. This need to organise and manage was typical of Dalhousie's methods and unsurprisingly soon after his arrival he recognised the importance of telegraphy to British India.

Dalhousie is considered by many scholars to be one of the most controversial figures in the history of India under British rule and although he has been blamed by many scholars as 'the major provoker of the Sepoy mutiny because of his policy of annexing one Indian State after another under the Doctrine of Lapse, he was nonetheless destined to go down in history as the author of certain famous reforms and industrial projects'.[37] Soon after his arrival in India, he wrote to a friend: 'The Government is 1200 miles distant from the war and with the post going at 4 miles an hour, is compelled to rely on its agents on the spot.' He had referred earlier in a letter of 3 June 1848

[34] Mel Gorman, 'Sir William O'Shaughnessy, Lord Dalhousie, and the Establishment of the Telegraph System in India', *Technology and Culture. The International Quarterly of the Society for the History of Technology*, 12, 1 (January 1971), 581–601, 584. Dalhousie later became a marquis as a reward for his achievements in India.

[35] Sir William Lee-Warner, *The Life of the Marquis of Dalhousie KT* (London, 1904), 2 vols, vol. 2, 191; Lee-Warner was an Indian administrator and author: F.H. Brown, revised by Kathleen Prior, 'Warner, Sir William Lee (1846–1914)', *Oxford Dictionary of National Biography* (2004).

[36] J.G.A. Baird (ed.), *Private Letters of the Marquis of Dalhousie* (London, 1910), 33.

[37] Krishna Lal Shridharani, *Story of the Indian Telegraphs. A Century of Progress* (New Delhi, 1954), 8.

118 AN INNOVATIVE PHYSICIAN AND SCIENTIST

to the Second Sikh War, where he described Moolraj in open revolt.[38] Rapid communications were essential if India was to remain part of the empire with Britain in full control.

The Court of Directors in a despatch to Dalhousie dated September 1849 wrote:

> The subject of establishing the electric telegraph in India is one of great importance and though it appears that such a means of communication would be highly advantageous to the state and to the Community many serious considerations are involved in the question... the expediency of establishing a system of Electric Telegraphy independently of those which may be made simultaneously with the construction of each railroad and in the event of your taking a favourable view of the subject, we should wish to be informed of the means which n your opinion could be best employed for carrying it out.[39]

To achieve these objectives Dalhousie set up a Department of Public Works, staffed initially by the Corps of Engineers with the proviso that an indigenous body of engineers, surveyors and overseers would in time be trained to carry out the planned works.[40] The intention was to drive forward a programme of works such as railways, the electric telegraph, canals and irrigation; the vast distances to be covered by rail and telegraphs was daunting, exacerbated by the fact that huge stretches of the country had never been adequately surveyed. However, the example of steam and railway development was encouraging where already in India steam power was changing travel with steamboats active on some rivers; from 1829, steamships were serving the route between Bombay and the Red Sea, and from 1831 steam power was increasing productivity in the new Calcutta mint. The drive to establish the latest technological developments more especially electric telegraphs in a country partly unexplored merely underlined the magnitude of the task faced by O'Shaughnessy, who was neither an engineer nor a surveyor. Despite the existence of a Corps of Engineers, it was a surgeon who was appointed to the role of superintendent of telegraphs simply because he was the man with experience.

Charles C. Adley, a civil engineer who was attached to the East India Railway Company, recalled that in 1850 he had sent an address to the chairman of the Court of Directors of the East India Company, 'wherein proposals were submitted for the establishment of a comprehensive system of political and mercantile lines throughout India...' Whether Adley through his railway

---

[38] Baird (ed.), *Private Letters of the Marquis of Dalhousie*, 29. Mulraj Chopra (1814–1851) led the uprising which led to the Second Sikh War.

[39] S.C. Ghosh, 94, from Dalhousie Papers, 215.

[40] J.A. Bridge, 'Sir William Brooke O'Shaughnessy', 103–120, 104.

'THAT MAN O'SHAUGHNESSY' AND ELECTRIC TELEGRAPHY

connections was aware that the telegraph lines in England generally ran alongside the railway track, thus giving him the idea for a telegraph network, cannot be confirmed but remains a possibility. Whether by coincidence or not, two months later reports were submitted to the government in India by Colonel Forbes and Dr O'Shaughnessy, possibly triggered by Adley's address to the Company's Court of Directors.[41] Later, Adley became a telegraph engineer and in 1853, as assistant to O'Shaughnessy, he took charge of a line of telegraph as superintendent.[42]

In India, the *Calcutta Review*, appraising the progress of the new technology, wrote how O'Shaughnessy 'who had been engaged for many years in similar experiments and had been successful in blowing up the wrecks of the *Equitable* and the *Sir Herbert Maddock*', sunken vessels obstructing river traffic on the approaches to Calcutta, was directed to prepare a report on the subject of electric telegraphy.[43] Dalhousie had initially approached the Military Board to report on the feasibility of developing a scheme of widespread electric telegraphy, presumably at this time being unaware of O'Shaughnessy's capabilities and experience; some months later, the Board replied having consulted both O'Shaughnessy and Col. Forbes, later Major General W.N. Forbes (1796–1856), a Scottish military engineer of the Bengal army. The Board in reply put forward the views of Dr O'Shaughnessy and Colonel Forbes in which the former advocated an aerial line and the latter a subterranean line. It was decided to test the merits of these proposals by constructing an experimental line between Calcutta and Diamond Harbour.[44]

When the Military Board of Bengal called for reports from Lieut.-Col. Forbes of the Bengal Engineers and from O'Shaughnessy himself, it was now that differences of opinion appeared. Whether these were real or manufactured because of O'Shaughnessy being seen to encroach on the army's territory and perhaps through jealousy, the friction was evident. O'Shaughnessy, although favouring the aerial route, admitted that it might prove impossible to establish on account of the 'mischief committed by birds and monkeys' and the danger arising for the exposure of so much metal to the thunderstorms which are of so frequent an occurrence in a tropical climate. The Military Board was asked to examined the proposals replying on 19 February 1850 with reports

[41] Colonel Forbes, later Maj-General Forbes William Forbes (1796–1855), initially with the Bengal Engineers, had wide experience of land surveying and canal construction; in 1836, he designed St Paul's Cathedral, Calcutta and was attached to the Calcutta mint.

[42] Charles C. Adley, *The Story of the Telegraph in India* (London, 1866), 3–6. Adley was somewhat critical of O'Shaughnessy and clearly was not an admirer.

[43] *Calcutta Review*, xv (January–June 1851), 215.

[44] S.C. Ghosh, *Dalhousie in India, 1848–56. A Study of His Social Policy as Governor-General* (New Delhi, 1975), 93, 94.

120 AN INNOVATIVE PHYSICIAN AND SCIENTIST

from both O'Shaughnessy and Forbes. Forbes and O'Shaughnessy disagreed as to whether the overground or underground system was the better choice, Forbes favouring the underground route, but in the end as an experiment it was decided that a line, partly underground and partly overground, be constructed, of which sixty-nine miles were overground and the rest subterranean at a cost of just over fifty-nine pounds per mile.

Eventually, on 1 April 1850, it was decided that the experimental line should run from Calcutta to Chunsura, exactly the route suggested by O'Shaughnessy, part to be underground and part aerial, and to be supervised by a committee comprising Dr O'Shaughnessy, Colonel Forbes, Captain Browne and Captain Thuillier. In June, Colonel Forbes said that in his opinion the experimental aerial route would infringe patent rights in England and that 'it was absolutely necessary that the construction of such a line should be entrusted to somebody who would be able to bring to the work the knowledge and experience derived from a practical acquaintance with the construction of similar works elsewhere'. This was clearly aimed at O'Shaughnessy, who could claim only his successful telegraph line of 1839–1840 as evidence of competence in construction projects. The Honourable Mr Bethune said that the colonel's fears were groundless as English patents did not apply in India. According to Ghosh, O'Shaughnessy concluded that an aerial telegraph would be at the mercy of destructive bird and monkeys and exposed to the thunderstorms and lightning strikes common in a tropical climate.

The Home Department, Calcutta, in April 1850, informed the Court of Directors that an experimental telegraph line was planned, receiving a positive reply in August at which point Dalhousie, enthusiastically endorsing the project, wrote:

> Notwithstanding the continued pressure of finance, I regard this matter of electric telegraphs as of such infinite moment in India that I recommend the sanction of Government to whatever sum may be necessary for conducting the experiment on a scale sufficiently large to enable those charged with it to carry on their labours with rapidity and the fullest efficiency.[45]

The outcome of their reports and disagreements was the construction of the experimental line of telegraph, half subterranean, half overground, thirty miles in length in the vicinity of Calcutta, to be supervised by O'Shaughnessy. That the decision to instruct O'Shaughnessy as the supervisor was not acceptable to Col Forbes is obvious and perhaps understandable –construction work was the preserve of army engineers not of surgeons. Nevertheless, the line was completed under the supervision of O'Shaughnessy between October and

---

[45] Sir William Lee Warner, *The Life of the Marquis of Dalhousie K T*, 2 vols (London, 1904), vol. 2, 192, quoting a letter from Dalhousie to the Court of Directors.

December 1851 and extended in March 1852, crossing the Rivers Hooghly and Huglee to reach the sea. The line from Calcutta to Diamond Harbour with intermediate stations at Moyapore and Bistopore became operational on 4 October 1851 and 'on 5 December the line was opened to the public, and to messages on banking business, legal matters and opium speculation'.

Later, O'Shaughnessy recalled his personal story of his first telegraph line:

> In April and May 1839 the first *long* [original italics] of telegraph ever constructed in any country was erected by the writer of these pages in the vicinity of Calcutta. The line was twenty-one miles in length, embracing 7,000 feet of river circuit. The experiments performed on this line removed all reasonable doubts regarding the practicability of working telegraphs enormous distances – a question then and for three years later, disputed by high authorities, and regarded generally with contemptuous scepticism.

In contrast to the experimental lines in Britain which ran beside railway tracks mostly in an urban space, O'Shaughnessy's original line, twenty-one miles in length, passed through extremely difficult terrain with rivers and water courses to navigate; eleven miles of iron wire or rod were attached to trees and to bamboo poles, the whole exercise being a considerable feat of engineering and organisation. Moreover, he knew from his own experiments the potential of water courses for conduction of electricity and put the theory to the test: 'embracing 7,000 feet of river circuit' in the Hooghly River and several thousand feet more using a separate canal. The line running from Calcutta to the sea was 'opened for official and public correspondence' and the timing of this could not have been better when, with the outbreak of hostilities in Burma in 1840, 'the services of the telegraph were thus brought into instant and practical requisition, and its incomparable capabilities tested with complete success'.[46]

This was a remarkable achievement, working in relative isolation in comparison to scientists in Europe and America such as Carl August von Steinheil (1801–1870), the German physicist, and Joseph Henry (1797–1878), the American scientist who knew of the potential of water as a conductor but had not built a telegraph line. Gorman points out that O'Shaughnessy was unaware of their work and therefore his system was entirely an 'independent invention', which he developed in the laboratory and then field tested.[47]

---

[46] W.B. O'Shaughnessy, *The Electric Telegraphy in British India – a Manual of Instructions for subordinate officers, artificers and signallers employed in the Department* (London, 1853), prefatory notice, 3.

[47] Gorman, 'Sir William O'Shaughnessy, Lord Dalhousie, and the Establishment of the Telegraph System in India', 583.

O'Shaughnessy had proposed to build the very first stretch of the first line from Calcutta to Chinsurah as he was convinced that:

[a] very large return would, in the opinion of the mercantile gentlemen consulted, be made from Mizrapore and mercantile and banking establishments of Muttra, and the Marwaree shroffs. The newspapers of Upper India would also contribute... I would consider it highly probable that these items with the amount above specified (Rs. 8 for 480 words for Calcutta and Bombay) would pay a large sum beyond the yearly expenditure and leave the telegraph eighteen hours available in the day for the use of the government without charge.[48]

The *Calcutta Review* of March 1851, in a highly detailed report of this technological advance, wrote:

The deepest interest, as might have been expected, was felt by all classes in this great national work; and information regarding its progress was eagerly sought by the conductors of the press and communicated from time to time to their constituents. From these successive records we learn that the work commenced on the 5th of November and was completed as far as Diamond Harbour, but that Dr. O'Shaughnessy was obliged to quit his labours on the 27th of January to take charge of the Assay office at the Mint: and this circumstance has prevented the extension of the line beyond thirty-two miles and a half. The wire, used by him, was an iron rod, three-eighths of an inch in diameter, coated with two layers of cloth, saturated with pitch, and then laid in a bed of roofing tiles, in a melted composition of three parts of sand and one of rosin, which, when cool, becomes as solid as a stone, and is impervious to white-ants, or vermin, or the saline influence of the soil. Before the completion of the experiment, the stock of rosin in the market was reduced, and the price rose to such an extent as to constrain Dr O'Shaughnessy to make the second section of his line with three layers of Madras cloth, saturated with pitch, and laid in the ground without cement. A considerable portion of the line to Diamond Harbour runs through a morass; and in many places the water was only kept out by baling. The line may therefore be considered not only subterranean, but subaqueous. As yet, the experiment has completely answered expectation, and messages have been signalled throughout with perfect ease; still, the undertaking is at present considered only in the light of an experiment, the result of which cannot be ascertained, with a view to ulterior operations, until it has been tested by a succession of thunderstorms, and by an entire rainy season. The greatest difficulty, which has been experienced, is in the instruction of a body of signallers. The class of native pupils was at once disbanded, on the death

[48] Choudhary, 'Beyond the Reach of Monkeys and Men?', quoting Home Department, Public Proceedings, 4 April 1850, No 48, Report 1, 348–9. Shroffs is a word describing Indian bankers or moneylenders.

'THAT MAN O'SHAUGHNESSY' AND ELECTRIC TELEGRAPHY

of one of them, at a little distance from Calcutta, from fever. They refused to leave town for an unhealthy locality, and it has been deemed necessary to place a class of European boys under tuition. Such is a brief narrative of the progress which has been made in this experiment, the full results of which will not be known till the commencement of the next cold season.[49]

Gorman writes admiringly of O'Shaughnessy having the boldness, the faith and the confidence in choosing 'such formidable conditions of construction and operation'. The line was in working order by March 1851 but modifications and the training of signallers delayed the opening of the line until October 1851, when shipping intelligence from Diamond Harbour to Calcutta was now rapid, the previous use of semaphores outdated. In February 1852, an extension was made to Kedgeree, fifty miles from Calcutta near the mouth of the Hooghly at a point where all ships sailing to and from the capital had to identify themselves.

O'Shaughnessy reported on the construction and working of the line from Calcutta to Kedgeree to J.P. Grant, secretary to the Bengal government, in which he described in detail the technicalities of the construction, taking care to explain that he [had] 'purposely selected this troublesome and objectional line [between the stations of Bishtopore and Moyapore] on the principle by which I have all through this undertaking been guided, [that is] to encounter the greatest difficulty first so as to know the worst'. He went on to describe the terrain as composed of rice swamps, jeels and creeks on which no road or embankment or bridge exists. Moreover, the work was done through the rainy season when the welding of the rods had to be carried out afloat in canoes. He emphasised that unlike in Europe or America no wire was used, going into detail as to the benefits of iron rods both in terms of cost, ease of use and ability to withstands the predations of birds and monkeys. Even the bamboo poles proved their worth during a severe storm when trees had been uprooted, houses of solid masonry destroyed, steamers driven ashore, many native ships destroyed but not one of the lines bamboo poles was damaged.[50] However, he admitted that 'the objections [he] recorded in 1850 to the European and American overground systems as advocated by Colonel Forbes remain totally

[49] *Calcutta Gazette*, March 1851.

[50] William Brooke O'Shaughnessy, *Selections from the Records of the Bengal Government number vii. Report on the Electric Telegraphic between Calcutta and Kedgeree* (Calcutta, 1852), 4. The report extends to twenty pages with appendices giving costs of construction, salaries and wages; J.P. Grant, later Sir John Peter Grant (1807–1893), was an administrator in Bengal, a freemason and a friend of O'Shaughnessy.

124 AN INNOVATIVE PHYSICIAN AND SCIENTIST

unaffected by the results now attained and described'. His method seemed to be efficient and more profitable than he had envisaged.[51]

O'Shaughnessy wrote in his monograph on the construction of the first line of electric telegraph in India:

> This line was commenced in October, 1851, and opened at Diamond Harbour in December of that year. In the following May a branch was led to Moyapore. In August and December it was extended to Kedgeree, 80 miles distance by the line followed; and in March, 1852, the rivers Hooghly and Huldce were crossed, and the line from Calcutta to the sea opened for official and public correspondence.

Dalhousie, in a minute of 14 April 1852, wrote:

> I have visited the line, and in common with hundreds of others can bear testimony to the beautiful simplicity of the work, to the regularity of its operations, and to the perfect success of it as a national experiment of the highest and most immediate moment to the interests of India, In truth, the best of all testimony is borne to it by the periodical delivery every three hours during each day of intelligence from each station between Calcutta and Kedgeree up to the hour at which the intelligence is delivered here in Calcutta.[52] Gorman considered that the line was 'unique in the history of telegraphy because it relied on a coded number of shocks transmitted from an induction coil for reception by the hands of an operator'.[53]

On 16 April 1853, the first railway ran a distance of 34 kilometres from Bombay to Thana, the same year which saw the introduction of the electric telegraph. The railway transformed the speed of communications as dramatically as did the telegraph, the latter being comparable to that of the internet and email in the modern era.

Fahie, in an early history of telegraphy, recounts how O'Shaughnessy, faced with the difficulty of crossing the rivers of India with his iron rods, adopted the idea of using subaqueous telegraphy, laying in 1849 an iron rod connected to batteries and instruments on each bank under the River Huldee which at this point was 4,200 feet across. He found that signals were actually transmitted but there were frequent interruptions and only skilled operators could use the system. A trial using the water alone as the sole conductor of the electric impulses produced intelligible signals, but the battery power required would never be feasible for practical purposes and would be prohibitively expensive.

---

[51] O'Shaughnessy, *Selections*, 8.

[52] W.B. O'Shaughnessy, *The Electric Telegraph in British India* (Calcutta, 1853), iv.

[53] Gorman, 'Sir William O'Shaughnessy, Lord Dalhousie, and the Establishment of the Telegraph System in India', 583.

However, he did not abandon the idea and in 1858 he experimented with the method on the lake at Ootacamund, observing:

> I have long since ascertained that two naked uncoated wires, kept a moderate distance-say 50 or 100 yards apart, will transmit electric currents to considerable distances (two to three miles) sufficiently powerful for signalling with needle instruments.[54]

He was certain that copper wire would last for many centuries even if exposed to the bare earth, 'as has been sufficiently proved by the condition of the copper plates (tambapatras) disinterred in India'.

The experiment was entirely successful, but the *Review* made clear that until it had been subjected to thunderstorms and a full rainy season future operations should not be undertaken. The article concluded saying that if the line had withstood every disturbance, 'it was to be hoped that Government will not hesitate to sanction the outlay necessary for carrying out the whole of Dr. O'Shaughnessy's plan'.[55]

Dalhousie's final minute to Parliament included this extract from O'Shaughnessy's report which highlighted the problems he faced and eventual overcame in the execution of his brief:

> Throughout Central India, for instance, Dr. O'Shaughnessy states, – 'The country crossed opposes [*sic*] enormous difficulties to the maintenance of any line. There is no metalled road; there are few bridges; the jungles also in many places are deadly for at least half the year; there is no police for the protection of the lines. From the loose black cotton soil of Malwa to the rocky wastes of Gwalior, and the precipices of the Sindwa Ghats, every variety of obstacles has to be encountered!'

As Gorman makes clear, the role of Dalhousie in pushing for a national network now became crucial, for in April 1852 he wrote to the Court of Directors stating 'everything, all the world over, moves faster nowadays... except the transactions of Indian business'. In order to expedite matters, he requested that O'Shaughnessy appear in person before the Court in London to promote the idea of a nationwide telegraph and at the same time he also asked them to approve an award of 20,000 rupees to O'Shaughnessy.

In the beginning of April 1852, the report of Dr W. O'Shaughnessy on the full completion and the successful working of the experimental line of electric telegraph, which had previously been authorized by the Honourable

[54] John Joseph Fahie, *A History of Wireless Telegraphy* (Edinburgh and London, 1901), 39–40.

[55] *Calcutta Review*, xv (January–June 1851), 219, 220. The description 'subterranean but subaqueous' was O'Shaughnessy's own.

126   AN INNOVATIVE PHYSICIAN AND SCIENTIST

Court, was laid before the government of Bengal. On the 14th of that month, the governor of Bengal strongly urged the governor-general in Council to obtain the sanction of the Honourable Court to the immediate construction of lines of electric telegraph from Calcutta to Agra, to Bombay, to Peshawar, and to Madras. He also advised that Dr O'Shaughnessy should be forthwith sent to England for the furtherance of the plan. On 23 April, the governor-general in Council recommended the proposed construction of the lines to the Court of Directors and agreed that Dr O'Shaughnessy depart for England. The Honourable Court entered into the proposal with the 'utmost cordiality and promptitude', and on 23 June it signified its assent to the whole proposal.

The *Calcutta Review* in June 1851 considered O'Shaughnessy's as yet unconfirmed report which was said to recommend that the lines should extend throughout India, from Calcutta to Agra, the latter to be the great centre of communication, to Simla and Lahore, to Bombay with a total of 2,500 miles.[56]

> The distance to be traversed is upwards of 3,000 miles, and it is intended to proceed with such expedition in its construction that its completion may be expected before three years from the present time. Dr. O'Shaughnessy has lately been employed in India in carrying on experiments which the electric telegraph, in order to discover the best system which could be adopted. The result of these experiments was highly satisfactory to the Governor-General and to the Court of Directors, who immediately resolved to take measures for giving to India the inestimable advantage of this marvellous means of communication.

On 14 April 1852, Lord Dalhousie, as governor-general, laid before the government of India 'a long and deeply-interesting Minute', in which his Lordship proposed the construction of lines from Calcutta to Agra, to Bombay, to Peshawar, and Madras; and the deputation of the author of this Manual to England, to give evidence Before the Court of Directors and assist in the dispatch to India of the requisite materials and stores.[57]

The favourable report on the experimental line between Calcutta and Kedgeree was decisive, with the result that Dr O'Shaughnessy was sent to England to present the proposal for telegraphic expansion to the Court of Directors. On 1 August, contracts were signed in London for the supply of 5,600 miles of wire. During the remaining months of 1852 and through the greater part of the next year, O'Shaughnessy was employed back in England

---

[56] *Calcutta Review*, xv (March 1851), 219.

[57] Suresh Chandra Ghosh, *Dalhousie in India 1848-56. A Study of his Social Policy as Governor-General* (New Delhi, 1975), 96. Ghosh quotes extensively from the Dalhousie Papers at the National Records of Scotland highlighting the vast extent of his correspondence.

procuring and despatching to India the immense quantity of materials which was required for the vast work which had been projected. The intention was to construct the lines as rapidly as possible and to achieve this bamboo poles were erected alongside the Great Trunk Road, the poles sited fifty feet apart and sunk three feet into the ground. Most of this preliminary work was carried out while the wire was being prepared in England under the supervision of O'Shaughnessy, so that by November 1853 the wire was ready to be placed on the bamboo poles and the construction of the telegraph line from Calcutta to Agra was commenced. Adley wrote that 'all the powerful resources of the Government were brought into play. The bullock train establishments, inland river steamers, commissariat and public works departments throughout the country, were more or less placed at the disposal of the telegraph...'[58]

On 24 March 1854, a message was sent over the line from Agra to Calcutta, 800 miles apart, the work having been completed in less than five months. The drive which was apparent at the commencement of this huge enterprise was maintained throughout all its subsequent progress. On 1 February 1855, fifteen months after the commencement of the work, O'Shaughnessy was able to notify the government of the opening of the lines from Calcutta to Agra and thence to Attock on the Indus and again from Agra to Bombay and thence to Madras. These lines included forty-one offices and extended over 3,000 miles. Dalhousie declared that this 'vast result' was due to the energy, ability and expertise of O'Shaughnessy.

However, there inevitably were problems, notably when a new line was opened to Ootacamund, where Dalhousie was residing, and he experienced delays in receiving messages, one from Bombay and one from Calcutta, both taking two days to arrive. He complained to the superintendent, who responded claiming that this was caused by the ignorance, the carelessness and the indifference of the signallers and his solution was to suggest the introduction of the Morse system. To enable this change to be introduced, he asked for permission to go to Europe and America and was again granted leave of absence until the end of 1856.[59]

When O'Shaughnessy returned to England in 1856, *The Times* recorded his presence as a guest at a celebratory dinner in honour of Professor Samuel Morse who was visiting London in October 1856. Morse made special mention of O'Shaughnessy's successful supervision of telegraphy in India, and in his response, Dr O'Shaughnessy intimated that 'he had made the journey from India to England to introduce into India the system of telegraph which had

---

[58] Roy Macleod and Deepak Kumar (eds), *Technology and the Raj: Western Technology and Technical Transfers to India 1700–1947* (New Delhi, 1995), 155; Adley, *The Story of the Telegraph in India*, 6.

[59] Ghosh, *Dalhousie in India 1848–56*, 105, 111.

128   AN INNOVATIVE PHYSICIAN AND SCIENTIST

been perfected by Professor Morse'.[60] In November, he was in attendance at Windsor Castle to receive his knighthood, an award which Dalhousie had promoted, writing from Southampton on 25 May 1856 to his friend Sir George Couper: 'I am now going to fight for O'Shaughnessy.'[61]

That there were teething problems in such an enormous enterprise is hardly surprising, but nonetheless *The Times* expressed its support:

> To Sir W. O'Shaughnessy undoubtedly belongs the merit of originating, organizing, developing, and superintending this great increment of our strength; and it would be in the highest degree unjust to detract from the reputation of a distinguished public servant who has reinforced every battalion in the field and informed every statesman and administrator in the closets by affixing to him personally the blame for faults which arise from matters over which he can exercise no supervision.[62]

The Times was not the only voice offering support, as was apparent when Dalhousie told the East India Company Board: 'that with the number of functionaries, boards, references, correspondences, and several Governments in India, what with the distance, the reference for further information made from England, the fresh correspondences made from that reference, and the consultation of the several authorities in England, the progress of any great public measure, even when all are equally disposed to promote it, is often delayed'[63]

Dodd, in his *History of the Indian Revolt,* wrote of the electric telegraph, describing it as the greatest invention of the age and finding in India 'a congenial place for its reception':

> Where the officials had no more rapid means of sending a message to a distance of a thousand miles than the fleetness of a corps of runners, it is no marvel that the achievements of the lightning-messenger were regarded with an eager eye. An experimental line was determined on, to be carried out by Dr (now Sir William) O'Shaughnessy; and when that energetic man made his report on the result in 1852, it was at once determined to commence arrangements for lines of immense length, to connect the widely separated cities of Calcutta, Madras, Bombay, and Peshawar, and the great towns between them.[64]

---

[60] *The Times*, 11 October 1856, p. 6, cols 5, 6.

[61] Dalhousie, Letters to Sir George Couper, 375.

[62] *The Times*, 2 April 1859, p. 12, cols 1, 2.

[63] Dalhousie to the Court of Directors, quoted by O'Shaughnessy.

[64] George Dodd, *The History of the Indian Revolt and the Expeditions to Persia, China and Japan 1856–7–8* (London, 1859), 9.

'THAT MAN O'SHAUGHNESSY' AND ELECTRIC TELEGRAPHY   129

Lieut.-Col. Sanders, in his two-volume work on military engineering in India, considered that India showed the world the value of field telegraphy in war.[65] The limited role of the military engineer in establishing telegraphy in India became a source of disagreement between the civilians and the army. The first military involvement in supervision of the telegraph line was not until May 1853 when Lieutenant Patrick Stewart of the Bengal Engineers was appointed superintendent in the absence of O'Shaughnessy in Europe. As acting super-intendent, Stewart oversaw the extension of the telegraph line by 1,280 miles to Lahore, with steamers and bullock carts the main means of transport. In July 1856, he was again appointed to act during O'Shaughnessy's absence and spent the following four months travelling between Calcutta and Allahabad, 'organizing the telegraph arrangements' during the rebellion until he handed over duties once more to O'Shaughnessy on 15 January 1858. During the Uprising of 1857–1858, the same Lieutenant Stewart, an experienced teleg-rapher by then, was of immense help to General Sir Colin Campbell as will be shown later.

Krishnalal Shridharani, a proud Indian Posts and Telegraphs employee writing in the volume commissioned to celebrate the centenary of the telegraph in India, was certain that the telegraph had brought the country together and was keen to stress that Indians had contributed to the early development under O'Shaughnessy. The length of the first line erected in 1851 may not have been impressive he wrote, but as he pointed out 'the difficulties surmounted were of gigantic proportions'. The low-lying delta of the Ganges was exposed to storms and electrical disturbance and underwater cables were liable to be damaged by anchors of ships and fishing boats:

> and yet nationhood in its modern sense that we have achieved is in no small measure due to the beginnings made in Calcutta by one Dr O'Shaughnessy, later to be supplemented by such Indian associates as Seebchunder Nundy who utilised such typical Indian materials as 'bamboo scissors' and toddy palm trunks when he present-day Hamilton poles were a far-off dream. Communications of any kind in India were in their infancy when Dr O'Shaughnessy started a little over 'the official century', to experiment with his telegraph wires. Even post was exchanged by couriers along few main roads connecting principal provincial towns with the seat of Government in Calcutta and they were 'reserved for official letters and parcels', only on rare occasions private individuals being given the privilege. It was in the Telegraph department that the first organised process of communication dawned in India.[66]

---

[65] E.W.C. Sanders, *The Military Engineer in India*, 2 vols (Chatham, 1935), 280, 283.

[66] Shridharani, *Story of the Indian Telegraphs*, 1; Shridharani used the spelling

130 AN INNOVATIVE PHYSICIAN AND SCIENTIST

Shridharani perhaps proudly and understandably made a point of mentioning the work of an Indian who was O'Shaughnessy's valued assistant, writing that according to a communication from the governor-general in Calcutta at least 'one Indian' had distinguished himself in the construction work of telegraph lines. He wrote admiringly of Seebchunder Nundy, the first Indian to make an official appearance in the records of the Telegraph Department and who was 'later to receive high praise both from the Government and the press and to become a Rai Bahadur'.

Nundy (1822–1903) was born in a poor family in Calcutta. In 1846, he joined the government service in the refinery department of the Calcutta mint under Dr O'Shaughnessy: 'The young Indian destined to be a telegraph pioneer was selected as personal Assistant by the great Irishman', and when in 1852 the Company decided to construct telegraph lines in India, Nundy was recruited by O'Shaughnessy and put in charge of the workshop. On completion of the experimental line, Nundy sent the first signal from the Diamond Harbour end which was received by O'Shaughnessy at the Calcutta Station in the presence of Lord Dalhousie himself. By the end of 1856, India had 4,250 miles of electric telegraph and forty-six receiving offices. During the Uprising of 1857, Nundy rendered excellent service, sometime acting as head of the Telegraph Department's headquarters.

In this next section the use of the telegraph during the rebellion of 1857 will be weighed up considering, on the one hand, the view that the telegraph saved India for the British, and, on the other, that it had little effect on the outcome of the war. Dalhousie, aided by the genius of O'Shaughnessy, had rapidly spread a network of electric wires 'across the whole length and breadth of the country', saying, 'it was a nice thing to do, a right thing to'. He had not foreseen the events that were to disturb the colonists in 1857 soon after his departure from India, some scholars attributing the violence to initiatives of Dalhousie himself in absorbing Indian kingdoms for the benefit of the Empire. The speed with which the Court of Directors in London agreed to the vast telegraphy scheme is surprising, perhaps revealing an apprehension that rebellious undercurrents amongst the Indian people were surfacing.

Arnold observed that 'it is an oft-repeated exaggeration to say that the electric telegraph saved India in the Mutiny of 1857, but they did carry to the authorities early word of the up-country revolt of May 1857', and of course as he points out one of the telegraph's first uses was to carry news of the

---

Nundy rather than Nandy, and that has been used here as the correct way. It is noteworthy that he used the word mutiny to describe the uprising.

fall of Rangoon in April 1852 to Dalhousie in Calcutta.[67] The exaggeration may well have been set in motion by an article in *Macmillan's Magazine* in October 1897 and later taken up by the *Daily News* under the headline 'How the Electric Telegraph Saved India.' The writer went on to claim that 'it is almost impossible to over-estimate the assistance which the telegraph renders, not only in the administration of the country, but in the conduct of every military operation that is undertaken'. The assistance of the telegraph to the administration of India was undoubted but that was of secondary importance during the military actions of 1857–1858. The writer observed: 'many have no doubt heard of the fateful telegram which led Robert Montgomery, the Judicial Commissioner of the Punjab, in reporting on the events of that anxious time to announce that "The electric telegraph has saved India"'; the writer recounted the heroic actions of William Brendish, at the time of publication (1897) the sole survivor of the Delhi telegraph office staff who was present in the office during the 'mutiny' and sent the fateful telegram from the Delhi telegraph office.

The notion has persisted into the twentieth century. A history of military engineering in India written in 1935 states categorically that 'the electric telegraph saved India in the Mutiny of 1857. It is difficult indeed to estimate what England owes to the persistence and courage of her first Superintendent of Telegraphs in Bengal.' The author goes on to quote Russell who wrote in *The Times*: 'never since its discovery has the electric telegraph played so important a role. It has served the Commander-in-Chief better than his right arm.' Russell concluded saying, 'thus India showed the world the value of field telegraphy in war'.[68]

Choudhury, writing of the 1857 uprising, says that telegrams sent by the signallers at Delhi warning of the rebellion were crucial to the British retaining their control over the Punjab, thus saving British power in India, an opinion shared by Sir John Lawrence, Chief Commissioner of the Punjab: 'The telegraph saved us', he claimed referring to the events in Meerut in May, when he was forewarned by the telegraph message sent from the office in Delhi.

Sir Robert Montgomery (1809–1887), judicial commissioner writing from Lahore on 18 August, was in no doubt as to the usefulness of the telegraph: 'Under Providence, the Electric Telegraph has saved India.' In praising the bravery of Charles Todd of the Delhi Telegraph Office, whose message led to the disarming of the native regiments in Lahore and Peshawar and the

---

[67] David Arnold, *Science, Technology and Medicine in Colonial India* (Cambridge, 2000), 114.

[68] Sanders, *The Military Engineer in India*, 279, 283. Sir William Howard Russell (1827–1907) was an Irish-born reporter for *The Times* who was perhaps the first war correspondent, reporting on both the Crimean War and the Indian War of 1857.

132 AN INNOVATIVE PHYSICIAN AND SCIENTIST

maintenance of the line from Delhi to the Punjab, Montgomery concluded that these officers 'rendered inestimable service to the Government of India and to the highest interests of the whole Empire'.[69] Fifty years later, a Mutiny Memorial to the Delhi Telegraph Office staff was unveiled in Delhi in 1902 with the following inscription:

> Erected on 19th April 1902 by members of the Telegraph Department to commemorate the loyalty and devoted services of the Delhi Telegraph Office Staff on the eventful 11th May 1857. On that day two young signallers William Brendish and I.W. Pilkington remained on duty till ordered to leave and by telegraphing to Umballa information of what was happening at Delhi rendered invaluable service to the Punjab Government. In the words of Sir Robert Montgomery 'The Electric Telegraph has saved India'.

There is no doubt that the brave signallers who continued at their posts sending messages while under attack alerted British officials in the Punjab, who were able to disarm rebellious sepoys in Meerut, an important action, but whether it saved India for the Empire is questionable.

Kaye, in his three-volume history of the events of 1857–1858, published twenty years after the uprising, does not write of the electric telegraph as a war-winning addition to the British defence of their colony, writing more about the local populace who viewed it as a sinister technology which had to be destroyed. He described how in January 1857, 'a few days after the story of the greased cartridges first transpired at Dum-Dum, the telegraph station at Barackpore was burnt down'. Earlier in the first volume, he wrote of 'the lightning post, which sent letters through the air and brought back answers… was a still greater marvel and a still greater disturbance [compared to the railways]'. Describing events at Allahabad in June 1857, Kaye reported that the railway works were destroyed, and the telegraphic wires were torn down. 'There seemed to be an especial rage against the Railway and the Telegraph.'[70] The events described by Kaye are largely unimportant episodes of vandalism rather than organised destruction.

Ghose discussing the role of the mutineers and the telegraph writes that 'as the telegraph was strictly under government control an overwhelming

---

[69] W. Coldstream (ed.), *Records of the Intelligence Department of the Government of the North-West Provinces of India during the Muting of 1857, Preserved by, and now arranged under the Superintendence of Sir William Muir, then in charge of the Intelligence Department* (Edinburgh, 1902), 469, 480. Dak is the Indian Postal Service; Adams, *Memoir of Surgeon-Major Sir William O'Shaughnessy Brooke*, 15, quoting Montgomery, the Judicial Commissioner.

[70] John William Kaye, *A History of the Sepoy War in India 1857–1858*, 3 vols (London, 1876), vol. 1, 193, 497; vol. 2, 257.

majority of signallers were of British or Anglo-Indian origin and most were loyal to the British'. Moreover, there was nobody on the rebel side who had any knowledge of signalling over what was called 'the lightning wire' and were unable to make use of the lines that fell into their hands – they could only destroy them. Ghose describes how 'with the spread of the mutiny to Cawnpore, Benares, and Allahabad in the first week of June, telegraph lines were destroyed whenever they fell into mutineers' hands, resulting in the isolation of Allahabad, Cawnpore, Lucknow and Agra'. In a study of Dalhousie's policy as governor-general, Ghosh writes that 'the westernising policy of reform and innovation reached its apogee during the administration of Dalhousie in India' with the railways, the electric telegraph and the post as 'the three great engines of social improvement', with the telegraph remaining in the hands of civilian 'engineers' rather than military men.[71] This is an important observation and confirms that there may well have been a conflict between civilians subordinate to O'Shaughnessy and army engineers who considered that the telegraph should be under their control.

O'Shaughnessy in his Telegraphic Report of 1856–1858 testified that during the uprising the whole line from Agra to Indore of 400 miles, Agra to Cawnpore of 180 miles and Agra to Delhi, 178 miles, had been destroyed, with the posts used for firewood and the wire cut for slugs or bullets or rendered perfectly unserviceable for telegraphic purposes. The first news of the uprising was flashed from Meerut on 10 May, simultaneously alerting the rebels to the danger from the wire and immediately the line between Meerut and Delhi–Agra was destroyed. Despite such evidence, O'Shaughnessy maintained that the telegraph was relatively immune from local attack, an opinion which was perhaps valid so long as the local rebels did not appreciate the danger above their heads.

The value of the telegraph to the military and the government is recounted by Shridharani, who quoted Lord Dalhousie, who highlighted two remarkable instances proving the telegraph's political value: when the 10th Hussars were ordered with all speed from Poona to the Crimea, Dalhousie received a message from the government of Bombay regarding their dispatch at 9 o'clock in the morning – a reply with instructions by telegraph was sent with an answer from Bombay received in the evening; a year before the orders for the speedy despatch of reinforcements to the seat of war could not have been made in less than thirty days. 'From thirty days to twelve hours was the telescoping of time by the telegraph and it made all the difference between and victory.'[72]

---

[71] Saroj Ghose, 'Commercial Needs and Military Necessities: The Telegraph in India', in Roy Macleod and Deepak Kumar (eds), *Technology and the Raj: Western Technology and Technical Transfers to India 1700–1947* (New Delhi, 1995), 166.

[72] Shridharani, *Story of the Indian Telegraphs*, 2.

The Records of the Intelligence Department, published forty-five years after the uprising, afford snapshots of the role of the telegraph during the uprising and not all are positive. On 12 August 1857, General Neill wrote to General Wilson from Cawnpore saying that the 'accounts from down country are better, although the dak between Benares and Calcutta is cut off and also the Telegraph. He said that the mutineers at Arrah 'got a severe dressing' but were still able to spoil 'the dak and the Telegraph'.

George Dodd's *History of the Indian Revolt* records how Sir Colin Campbell carried the electric telegraph with him for camp to camp and from post to post, this made possible by the efforts of Lieutenant Patrick Stewart, who set up poles and wires extended to wherever Sir Colin went: Allahabad, Cawnpore, Buntara and the Alum Bagh:

> No sooner did Sir Colin advance a few miles, than Stewart followed him with poles and wires, galvanic batteries and signalling apparatus... it may almost literally be said that wherever he lay down his head at night, Sir Colin could touch a handle and converse with Lord Canning. The value of the electric telegraph was quite beyond all estimate during these wars and movements: it was worth a large army in itself.[73]

Dodd relates how on 7 February 1856, as soon as 'the administration of Oude was assuredly under British government, a branch-electric telegraph from Cawnpore to Lucknow was forthwith commenced; in eighteen working days it was completed, including the laying of a cable, six thousand feet in length across the river Ganges'. Soon after Lord Canning's arrival as governor-general, a telegram to Oude enquiring if all was well was sent in the morning and by noon a reply saying 'all is well in Oude' was received.

As Wenzlhuemer points out, there are two sides to the story: 'the imperial narrative popularised by officials of the Empire and spread by the press, the tale of how the telegraph saved India but on the other hand most of the important nodes in the network were easily put out of operation and the system proved to be vulnerable to the point of uselessness'.[74] Nevertheless, despite the many claims of the importance of telegraphy quoted above and undoubted episodes of bravery, the evidence on balance suggests that British propaganda obscured the fact that the telegraphic system was often inefficient, badly designed and frequently not fit for purpose.

1857 was a communication crisis of enormous proportions for the British in India, and the role of the telegraph has generated many myths around

---

[73] Dodd, *The History of the Indian Revolt*, 11.

[74] Roland Wenzlhuemer, *Connecting the Nineteenth-Century World. The Telegraph and Globalization* (Cambridge, 2013), 211. Wenzlhuemer points out that Montgomery meant that the telegraph had saved India for the British rather than from the British.

the uprisings. The centenary book wisely avoided discussion of the role of the telegraph in the uprising of 1857–1858 preferring to highlight how better communications cemented a feeling of nationhood. The admiration and fondness for O'Shaughnessy and his excellent relationship with Nundy, his assistant, are expressed without reservation one hundred years after the event, a remarkable testimonial.

CONCLUSION

Historiography has not served O'Shaughnessy well, hence the writing of this book to address the deficit. The story recounted here of the remarkable career of Sir William Brooke O'Shaughnessy has shown how a man from Limerick in Ireland became a distinguished scientist and physician in both Britain and India. His achievements in the fields of pathophysiology and pharmacology alone would have been enough to place him among the pantheon of great medical innovators. Curiously, however, it was not for his successes in medical innovations that he was knighted by Queen Victoria in 1856, but for service to the Empire having successfully introduced electric telegraphy to India. These achievements and accolades should have been a guarantee of acceptance and appreciation by his peers and by scientists of his generation and those coming later. It is true that he was appointed a fellow by the Royal Society in 1843 but that distinction and his knighthood apart he has not received the recognition that his successes might have merited.

His obituary in the *British Medical Journal* illustrates very well just how little his medical research and publications seemed to matter to his colleagues in the medical profession. This journal, by 1889 perhaps the most widely read and respected medical publication in Britain, wrote: 'He was the author of numerous works on scientific and engineering subjects, particularly one on the electric telegraph published in Calcutta in 1839.'[1] There was no mention of his ground-breaking analysis of the blood of cholera victims carried out in the north of England and published in *The Lancet* during the first epidemic to visit Europe in 1831. This undoubted milestone in infectious disease pathology and therapeutics was and still is largely ignored by medical and social historians. The same fate has befallen his careful work on the efficacy of cannabis in a variety of disorders and his introduction of Indian hemp as a new medication to the West. The reasons for this are complex. The fact that around 1860 he 'disappeared', having changed his name, becoming Sir William Brooke for reasons which will be explained later, may have had some influence on this.

A position as assistant surgeon in the service of the East India Company

---

[1]   'Obituary of Sir William Brooke O'Shaughnessy', *British Medical Journal*, 19 January 1889.

when looked at from the standpoint of the twenty-first century does not appear an auspicious career move for an ambitious young doctor, such as O'Shaughnessy. The Bengal medical service of the East India Company was established in 1763 and, according to Arnold, it seems that the social and professional status of medical officers was never high, and in the Company before 1857 'they were looked down upon by the civilian and military elite'. These social and medical snobberies apart there were other more serious hardships to face: they did not always survive the rigours of tropical diseases and the Indian climate: one-third of all Company doctors between 1764 and 1838 died during their period of service.[2] O'Shaughnessy's eventual solo editing of the *Bengal Pharmacopoeia* is testament to this sad statistic, many of the editorial committee succumbing to disease before completion of the volume. Even as late as the 1830s, Company surgeons seemed to be treated with low esteem in England if a report in a medical journal is to be believed: 'the medical practitioner, in the service of our Honourable East India Company, is estimated somewhat under a *butler* in London! By the said Company a man is considered as far inferior to a horse – and consequently a surgeon is subordinate to a *black-smith!*'[3]

The Company medical service may not have been a highly attractive career option but for the young O'Shaughnessy, married with a child, who had found doors closed to him in London, unable to establish himself in the city through no fault of his own, it may have been the only solution. However, while still in London he had not given up hope and attempted to become established in forensic pathology and toxicology when in 1831 he applied for the chair of medical jurisprudence in the University of London at Kings College but was unsuccessful. Another endeavour was less personal but equally important in the eyes of someone rejected through prejudice and jealousy: in association with Thomas Wakley, founder in 1823 of *The Lancet*, he became involved in 1831 in an action group which was trying to establish a London medical college, becoming secretary with Joseph Hume MP as chair. Hume had served in the East India Company medical service and his success in India may have encouraged O'Shaughnessy to follow the same path.

When the cholera epidemic arrived in England, O'Shaughnessy was asked to go to Sunderland where the analysis he carried out on cholera victims in the north-east, based entirely on scientific principles, was exceptional at a time when medicine was still viewing disease in Galenic terms in which humoral theory was pre-eminent. Although there was a minor move away from

---

[2]   David Arnold, *Science, Technology and Medicine in Colonial India*, series: The New Cambridge History of India (Cambridge, 2000), 58, 61.

[3]   282 Anon, 'Review of the Medical Department of the East India Company', *Medical-Chirurgical Review*, 13, 25 (1 July 1831), 112–122.

CONCLUSION

rigid humoral theory when William Cullen, Edinburgh professor of medicine, introduced a different theoretical classification based on the supremacy of the nervous system, nevertheless Cullen's new theories as to disease aetiology did not entirely replace the four humours. This concept remained the rationale, if such a term can be used, for a theory which from a twenty-first-century viewpoint was devoid of logic, a theory which drove blood-letting as a treatment. The humoral logic was that the thick tarry blood was the cause of cholera and had to be removed at all costs. The length to which physicians went to remove this diseased blood was extraordinary: at times when venous extraction failed, they turned to arterial puncture.

The question must be asked as to why physicians persisted in their advocacy of blood-letting as a valid treatment when it appeared that, apart from transient improvement, the success rate was nothing short of disastrous. There is no logical physiological explanation for the brief improvement they saw and recorded except possibly for the place of such a dramatic intervention. This idea has been advanced as a possibility by one physician who considers that the placebo effect should not be underestimated, but whether such an outcome was likely in cholera with a patient in a desperate state of hypovolaemic shock is doubtful and he made his claim discussing other conditions not cholera.[4]

The return from Paris of James Craufurd Gregory, the son of Professor James Gregory, changed the dynamic of medical progress in Edinburgh. Gregory graduated MD Edinburgh in 1824 and thereafter spent three years with Laennec in Paris, returning in 1827. Gregory was appointed a physician to the Royal Infirmary, Fellow of the Royal College of Physicians and the Royal Society of Edinburgh soon after his return and when the Edinburgh Board of Health was formed in 1831 to combat cholera in the city he became secretary. How O'Shaughnessy, newly qualified as a doctor, looked upon these distinctions bestowed on a colleague whose ideas about disease to an extent were rooted in ancient theory is not recorded, but subsequent developments must have disturbed him. Gregory's rapid acceptance in Edinburgh undoubtedly owed much to his family background, as son of Professor James Gregory, professor of the practice of medicine in Edinburgh who succeeded William Cullen in the chair, and as the scion of a Scottish medical dynasty, he was exceedingly well connected. His family background encompasses both Edinburgh medical aristocracy and Highland landed gentry, the latter in the shape of Donald Macleod of Geanies in Easter Ross whose daughter Isabella was James Craufurd Gregory's mother. Macleod, whose son James Craufurd Macleod (1775–1821) was part-owner of the plantation Geanies in

---

[4] D.P. Thomas, 'The Demise of Bloodletting', *Journal of the Royal College of Physicians of Edinburgh*, 44 (2014), 72–77.

140   AN INNOVATIVE PHYSICIAN AND SCIENTIST

Berbice, Guyana, was a landowner with a reputation for dispossessing tenants and crofters on his estate in Ross-shire.

Soon after his return to Scotland, Gregory published in 1829 a revised edition of Cullen's *First Lines on the Practice of Physic*, based on the latest edition of that work published in 1789 with Gregory's own amendments and additions.[5] This seemingly revisionist book, based on theory now forty years old, and his subsequent rapid promotion, could not have endeared him to the more progressive practitioners and students in the medical school, one of whom at this time was O'Shaughnessy. Gregory's rapid promotion may well have convinced him that without suitable contacts and patrons he had no future in Edinburgh.

The difference in understanding of disease causation between the two men is stark: Gregory, in the appendix to this new edition of Cullen, wrote that cholera was caused by 'an obvious affection of the nervous system... also an uncommonly great and sudden alteration of the circulation and distribution of the blood'. Gregory had no experience of cholera but recommended treatment with opium and wine or brandy, calomel and early blood-letting, while admitting that a flow of blood was often difficult to achieve.[6] It must have been a step too far for a traditionalist like Gregory to accept that intravenous infusion was preferable to blood-letting, one of the accepted treatments at this time. As the physician in charge of the Queensberry House Cholera Hospital, one of three in the city, and as secretary to the Edinburgh Board of Health, he had considerable influence. It seems that he used this to decry Dr Thomas Latta's implementation of O'Shaughnessy's recommendation to use intravenous saline. It was alleged by Dr Lewins, Latta's fellow practitioner in Leith, that Gregory was one of the Edinburgh physicians very much opposed to the use of intravenous saline, opposition from such an influential Edinburgh physician carrying great weight. Latta, a mere Leith doctor, could never prevail. O'Shaughnessy in London was in no position to change the views of Edinburgh physicians. However, Robert Christison, at the behest of the Dutch government, looked into the method and was not opposed to it, saying that if he was treating patients with cholera, he would certainly try it.[7] Christison was a member of the Edinburgh Board of Health along with other Edinburgh physicians; in their second report dated 26 January 1832, they made no reference to O'Shaughnessy's dramatic new findings and recommendations which were communicated to the London Board on 7 January 1832.

---

5   William Cullen's *First Lines on the Practice of Physic*, commenced by William Cullen MD completed by James Crawford Gregory (Edinburgh, 1829).

6   Gregory, in *First Lines on the Practice of Physic*, 375.

7   Robert Christison, 'Paper drawn up for the Guidance of the Dutch Government', *London Medical Gazette*, 10 (1832), 451.

CONCLUSION

The Board was enthusiastic about the value of blood-letting in various stages of the disease and in advice reminiscent of recent Covid measures, they told the public to avoid gatherings of any kind to prevent spread. James Craufurd Gregory died from typhus fever in 1832.

The epidemic ended in late 1832 with the inevitable result that interest in cholera largely evaporated together with any discussion about this new treatment. *The Scotsman* newspaper avoided in June 1832 publishing Latta's paper on saline injections, claiming there were reasons for this, but no explanations were made. The question arises as to whether they were influenced by the Edinburgh Board of Health, who were known to have reservations about this new therapy.

O'Shaughnessy's departure to India and the death of Latta from pulmonary tuberculosis in 1834 meant that two voices were silenced, one forever. Latta's obituary in an Edinburgh newspaper, the *Caledonian Mercury*, recalled his labours the previous year: 'Dr Latta was favourably known to the profession by the boldness and resolution with which he introduced and practised the injection of saline solutions into the veins in malignant cholera.'[8]

The death in 1837 of Dr James Mackintosh, a colleague of Latta's in the Drummond Street Cholera Hospital, an Edinburgh lecturer in medicine and a fervent supporter of Latta and the treatment with saline, was another blow to the work of O'Shaughnessy and Latta. Undoubtedly, these have all been factors in the lack of interest in O'Shaughnessy by medical historians and social historians alike. The latter group are generally less interested in a single medical discovery than in the social context of the discovery and, although unquestionably cholera has been a fertile source of material both for historians and for politicians, this medical advance has not excited much interest. Until the cholera vibrio was identified by Robert Koch in 1884, most observers and writers concentrated on how the disease was transmitted – on its epidemiology and the social conditions that they alleged promoted transmission. Edwin Chadwick, who was a member of the Board of Health during the second British cholera epidemic of 1848–1849, typifies the social activist whose enthusiasm for sanitary reform had earlier in the decade led to *The Report on the Sanitary Condition of the Labouring Population of Great Britain in 1842*. In Edinburgh, William Pulteney Alison, Professor of Medicine was rightly convinced of the appalling effects of poverty on health and on transmissible disease and took no part in the intravenous saline controversy. A single episode of scientific brilliance could not compete with the evidence of urban squalor collected and promoted by Chadwick and his colleagues. The persistent belief in miasmas as the prime cause of so many diseases was not altered in any way by O'Shaughnessy's analysis.

---

[8] *Caledonian Mercury*, 2 December 1833, p. 4, col. 3.

142   AN INNOVATIVE PHYSICIAN AND SCIENTIST

It is surprising that O'Shaughnessy showed little or no interest in cholera in India although it was well-nigh endemic during his time there. There was one occasion only when his work in Sunderland was discussed at some length but there was no further research into intravenous saline and thus the reputation of the man who originally recommended the treatment was never enhanced. David Arnold, in an essay on cholera in India, writes of:

> The seemingly arbitrary way in which the disease chose its victims, the rapidity of its advance, the high fatality among those attacked and the violent nature of their death all contributed to making cholera in India, as in the west, a singularly frightening disease, and one amenable to few customary controls.[9]

It is intriguing that O'Shaughnessy largely ignored cholera in Bengal perhaps seeing the impossibility of controlling the disease by the kind of measures used in Britain in 1831–1832 and that his recommended treatment of intravenous saline could never be implemented; the use of intravenous saline as used by Dr Latta in Edinburgh would not be safe or tolerated in rural Bengal. His pragmatism led him to concentrate on matters he could control.

In his several roles in Bengal, as professor of chemistry and *materia medica*, as a member of the Asiatic Society, scientific research projects, with many publications, it must be emphasised that he was not in a centre of excellence in Britain where his work would have counted for more. His cholera analysis and conclusions deserve to be recognised as milestones in medicine and therapeutics together with his colleague Dr Thomas Latta. Perhaps the last word should be left to Robert Pollitzer, whose book on cholera for the World Health Organization published in 1959 included these words: 'by common consent Latta is accorded the credit of having initiated the method of infusion treatment, of which ample and most beneficial use is still being made'.[10]

As a spinoff from his editorship of the *Bengal Dispensatory and Pharmacopoeia*, O'Shaughnessy became interested in the properties of cannabis, or bhang as it was known in Bengal. Bhang was recognised as one of the five sacred herbs with the properties to allay anxiety as noted in Ayurvedic texts in the second millennium BCE.[11] By careful testing of the drug first on animals, then on humans and finally, when satisfied of its potential and

---

[9] David Arnold, 'Cholera and Colonialism in British India', *Past & Present*, 113 (1996), 118–151, 119.

[10] Robert Pollitzer, *Cholera* (Geneva, 1959), 781.

[11] Ethan Russo, 'History of Cannabis as a Medicine', in Geoffrey Guy, Brian Whittle and Philip Robson (eds), *The Medicinal Uses of Cannabis and Cannabinoids* (London and Chicago, IL, 2004), 1–16, 1.

CONCLUSION 143

safety, he prescribed it with success for several hitherto untreatable conditions including tetanus, hydrophobia and other convulsive diseases. His attention to detail is evident when he thought to include the recipe for the tincture of hemp that he had himself used to treat patients.

As discussed in the chapter on cannabis, there were successes and failures when the drug was used in the West, most due to a variability in the plant's tendency to deteriorate over time and an inevitable imprecision in strength according to the plant's age and the form the pharmacist made up. In the end, these were found to be unacceptable to a profession increasingly science-based with dosage accuracy and precision an imperative. But as several scholars have pointed out, there were other factors involved in the disappearance of cannabis from the world's pharmacopoeia. In the USA, Lester Grinspoon, a Harvard professor of psychiatry, wrote in 1994: 'the government's commitment to gross exaggeration of the harmfulness of cannabis has made it necessary to deny the drug's medical usefulness in the face of overwhelming evidence'. Grinspoon had moved from being a believer in the harmfulness of recreational use of cannabis to an acceptance that most of the evidence was exaggerated. Grinspoon also pointed out that between 1839 and 1900 more than one hundred articles appeared in scientific journals describing the medicinal properties of the plant.[12] There has been in recent decades renewed interest in the therapeutic uses of cannabis based on medical judgements not on prejudice against its recreational use. Unlike his work on cholera, O'Shaughnessy and cannabis remain known and admired.

The Royal Society of London elected him a fellow recognising the importance of his scientific work including both his cholera analysis and his research into cannabis. Several prominent medical men were among his sponsors, all aware of his achievements in the field of medicine, but thereafter his role as supervisor of electric telegraphy in India overshadowed his successes in pathology and pharmacology to the detriment of his reputation in medical history.

O'Shaughnessy became a freemason in Calcutta on 23 May 1844, perhaps an unusual step for a Catholic Irishman, but one historian of freemasonry has highlighted its importance in the Empire:

> Freemasonry took an active part in displaying, promoting, and celebrating the Empire. Its universal rhetoric offered an inclusive idiom that could portray British presence in India as a consensual partnership while showcasing India as a timeless and unchanging society. When this illusion of permanence was shaken by the Indian Revolt of 1857, its rhetoric of universal brotherhood, highly syncretic ritual, and close ties with the British monarchy

---

[12] Lester Grinspoon, *Marihuana Reconsidered* (Cambridge, MA, 1994), 11, 15.

144   AN INNOVATIVE PHYSICIAN AND SCIENTIST

were perceived as valuable assets in the attempt to place the Indian princely aristocracy under the authority of Queen Victoria. All in all, colonial India was the prime site on which the British rehearsed the show. It was home to a cult of Empire bolstered and circulated by Masonic lodges.[13]

His admiration for his Bengal colleague, John P. Grant, a freemason, evinced in a letter to Hayman Wilson, perhaps explains his decision.

The third of O'Shaughnessy's innovations was in the field of electric telegraphy and it is here we realise the extent of O'Shaughnessy's versatility. At the same time, 1831–1835, that Joseph Henry was experimenting in the USA with electromagnetism, O'Shaughnessy was in London investigating the pathophysiology of cholera, before shortly joining the East India Company as an assistant surgeon. Soon thereafter, he became the first professor of chemistry in the new Calcutta Medical College and was publishing his results on the medical uses of cannabis. But by then, he had already begun experimenting with electric telegraphy and although his research was interrupted by his return to England on medical furlough in 1841, he had shown that electric telegraphy was indeed practical.[14] He had demonstrated by carrying underwater signals in the River Hooghly in Calcutta in 1838 that this method was possible, the first example in the world of underwater telegraphy.

The arrival of Dalhousie as governor-general was the catalyst for change: he ordered the experimental line to be constructed to Diamond Head in October 1851 and onwards to Kedgeree later that year. The line was finished in 1852 and was an immediate success proving its value commercially and perhaps more significantly a military success during the war with Burma. O'Shaughnessy wrote:

> Those results, having been duly reported, were under the consideration of the Supremo Government of India when hostilities commenced in Burma. The services, of the telegraph were thus brought into instant and practical requisition, as incomparable capabilities tested with complete success. The "Rattler," steam-frigate, bringing intelligence of the first operations of the war, had not passed he flagstaff of Kedgeree, on the 10th of April, when the news of the storming and capture of Rangoon was placed in the hands of the Governor-General in Calcutta, and posted on the gates of the Telegraph Office, for the information of the public.[15]

---

[13] Simon Deschamps, 'Masonic Ritual and the Display of Empire in 19th-Century India and Beyond: Showcasing Empire, Then and Now: Material Culture, Propaganda and the Imperial Project', *Printemps* (2021), https://journals.openedition.org/cve/8990 (accessed 8 July 2024).

[14] Choudhary, 'Beyond the Reach of Monkeys and Men?', 259–238, 339.

[15] O'Shaughnessy, *The Electric Telegraph in British India – A Manual of Instructions*

CONCLUSION 145

Whether the telegraph influenced the outcome of the 1857 rebellion in any way is controversial. As Wenzlhuemer rightly points out there are 'two sides to the story of the telegraph in the uprising. There is the imperial narrative popularised by colonial officials and spread by newspapers in Britain and the colonies. This is the tale of how the telegraph has saved India.' On the other hand, he claims the 'system proved to be vulnerable to the point of uselessness'.

In 1854, reviewing O'Shaughnessy's work supervising the electric telegraph in India, the writer in *The Lancet* concluded:

> Of Dr O'Shaughnessy himself we can speak in the highest terms of praise. Twenty years ago, he was one of the staff of the *Lancet*, and often have we had reason to be proud of the great ability that he displayed. Had he remained in Europe, not any man could have surpassed him in the department of science to which the then devoted his extraordinary powers of mind. He is a man of honour and a gentleman; and the Indian government may exult in having found a man endowed with such unusual qualifications for carrying into execution a scheme of operations which would reflect credit on the most enlightened government in the universe. Long may Dr O'Shaughnessy live to witness the triumph of his sagacity and profoundly scientific labours.[16]

After a long meeting with Lord Palmerston, the prime minister, Sir William left London for India on 1 October 1857, travelling via Constantinople where he met with officials of the Turkish government. The discussion centred around the development of a telegraph line from the capital to Baghdad, a line which he was to construct and supervise. It was intended that the line connected with the East India Company's line through the Persian Gulf to Kurrachee.[17] For whatever reason, O'Shaughnessy was never involved as will be discussed in the epilogue.

The Medical Reporter in 1895, in a series entitled 'Indian Medical Celebrities', wrote the following about O'Shaughnessy:

> It is strange that so little has been done to keep his memory green in the seat of his labours. Except the bust which is placed in an obscure corner of the Central Telegraph Office at Calcutta, the Medical College, in the estab-lishment of which he played such a conspicuous part and the reputation of

---

for subordinate officers, artificers, and signallers employed in the Department, iv, v.

[16] 'Reviews and Notices of Books', *The Lancet* (18 February 1854), 189. It is possible that his erstwhile colleague Thomas Wakley penned this tribute.

[17] *The Homeward Mail*, 16 October 1857, p. 46, cols 4, 5, 6, recorded the departure of Sir William en route to Turkey for discussions with the Turkish government following a lengthy meeting with Lord Palmerston.

which was not a little due to the researchs [*sic*] which he prosecuted within its wall, knows him not except by name. No statue, no bust, not even a portrait of the first professor of chemistry whose works about a new era in the history of Indian materia medica, graces the walls of the great institution of medical learning in India.[18]

[18] B.D. Basu, 'Indian Medical Celebrities', *The Medical Reporter*, v (January–June 1895) (Calcutta), 205.

EPILOGUE

After a lengthy meeting with Lord Palmerston, the prime minister, Sir William left London for India on 1 October 1857, travelling via Constantinople where he met with officials of the Turkish government. The discussion centred around the development of a telegraph line from the capital to Baghdad, a line which apparently he was to construct and supervise. It was intended that this line connected eventually with the East India Company's line through the Persian Gulf to Kurrachee. His involvement with the project did not progress. There may have been more than one reason for this: the Turkish authorities were not prepared to cooperate; a decision was made by the British government to develop the line using military engineers with Lieut.-Col. Patrick Stewart in control; finally, domestic troubles in O'Shaughnessy's life may have prevented him from superintending the project. Stewart had overseen a section of the telegraph line in India when O'Shaughnessy was back in London recruiting experienced telegraphers. His success there was sufficiently impressive that in 1863 he oversaw the cable laying from India to the head of the Persian Gulf where it was to link with the Ottoman government's land cable. The impression left at the time is that an army engineer, younger and equally experienced in telegraphy, was to take over and that O'Shaughnessy was side-lined.

Of more immediate concern to Sir William was the state of his marriage. He and his second wife, Margaret, were estranged – in February 1858, his wife and three children arrived back in Southampton having travelled from Bombay on the ship *Madras.* To all intents and purposes, this was the end of the marriage, but a divorce was never finalised, and Sir William was not free to marry again until Margaret died in June 1870.

Surgeon-Major Sir William Brooke O'Shaughnessy left India for the last time on 13 June 1860, travelling back to England on medical furlough for eighteen months, finally retiring from the Bengal service in 1862 after close on thirty years' service. From that time until his death on 8 January 1889, he lived in relative obscurity, an unusual situation for a man whose life had been dedicated to science and whose impact in medicine, pharmacology and telegraphy had been outstanding and very public.

In 1861, he changed his name by royal decree to O'Shaughnessy Brooke, at the time living in an illicit relationship with Julia Greenly Sabine. Presumably, the name change was an effort to conceal his domestic arrangements and to some extent protect Julia and their children. In Victorian times, living with a woman out of wedlock was a social catastrophe and by changing his name from O'Shaughnessy to Brooke he hoped to hide his perceived immorality from his peers and colleagues. For a time, they lived in Brighton where a son was born, until the family moved to Paris, where known as Sir William Brooke he lived a life of anonymity. Another child was born and from Paris they moved to Wiesbaden where a third child was born.

One week after his second wife died, William and Julia travelled to Edinburgh to arrange an 'irregular marriage', which although perfectly legitimate under Scottish marriage laws, was irregular in the sense that it was not a marriage solemnised in church and conducted by a clergyman. However, an irregular marriage could be registered if the couple presented themselves before a sheriff or magistrate and this is exactly what happened. The Scottish records show that on 18 June 1870 Sir William O'Shaughnessy Brooke, age sixty, widower, late of Calcutta, married Julia Greenly Sabine, age forty-one, by declaration in the presence of Robert Landale, Solicitor, sanctioned by Warrant of the Sheriff-Substitute of Edinburgh. Incidentally, both partners falsified their ages by two years to appear younger. Thus, it came about that of his three marriages, two were celebrated in Edinburgh. The forty years and more that had passed between the arrival of a young Irish medical student in Edinburgh and his third marriage in the same city of a knight of the realm were remarkable for a career full of innovations. Thereafter, his remaining years were passed living quietly in Hampshire.

BIBLIOGRAPHY

Manuscript and Archival Collections

British Library
Asian and African Studies
Ship Catherine: Deck Log 15 July 1833–17 June 1834.
IOR/L/MAR/B/97C.
H.H. Wilson Collection: MSS.EUR.E.301. Correspondence of O'Shaughnessy
to H.H. Wilson.

Royal Society
Correspondence of John Frederick William Herschel: 1812–1869: HS3.234; HS3
235 HS3 236 HS/13.134 HS 13/135. HS13 186.
Letter from Sir William O'Shaughnessy Brooke, 5 June 1864 MC/7/88.

Royal Asiatic Society of London
Minutes of General Meetings, December 1839 to March 1845.

National Records of Scotland
Dalhousie Papers, GD 45/6/215.

Trinity College Dublin
Medical School Register (TCD MS 758).

University of Edinburgh
Medical Matriculation Index 1783–1968.
Matriculation Albums 1627–1980 EUA IN1/ADS/STA/2.
Matriculation Indexes and Class Lists 1783–1986.

Wellcome Collection
*Papers relative to the disease called cholera spasmodica in India, now prevailing
in the north of Europe: with letters, reports and communications received from
the continent* (London, August 1831).

Primary Sources

Contemporary Books and Articles

Adams, M., *Memoir of Surgeon-Major Sir W. O'Shaughnessy Brooke, Kt, MD,*

150 BIBLIOGRAPHY

FRS, FRCS, FSA in Connection with the Early History of the Telegraph in India (Government Central Printing Office, Simla, 1889), facsimile edition.

Adley, Charles C., *The Story of the Telegraph in India* (London, 1866).

Ainslie, Whitelaw, *Observations on the Cholera Morbus of India. A Letter addressed to the Honourable the Court of Directors of the East India Company* (London, 1825).

— *Materia Indica*, 2 vols (London, 1826).

Arnold, Edwin, *The Marquis of Dalhousie's Administration of British India*, 2 vols (London, 1865).

Bell, George Hamilton, *Treatise on the Cholera Asphyxia* (Edinburgh, 1832).

Christie, Alexander Turnbull, *Observations on the Nature and Treatment of Cholera and on the Pathology of Mucous Membranes* (Edinburgh, 1828).

— 'A Treatise on the Epidemic Cholera; containing its History, Symptoms, Autopsy, Etiology, Causes and Treatment', *London Medical and Physical Journal*, May 1833, 83, new series, 353–365.

Christison, Alexander, 'On the Natural History, Action, and Uses of Indian Hemp', *Monthly Journal of Medical Science* (1851), 117–121.

— 'On the Natural History, Action, and Uses of Indian Hemp', *Edinburgh Medical Journal* (1851), 26–45.

Christison, Robert, *A Treatise on Poisons, in relation to Medical Jurisprudence, Physiology, and the Practice of Physic* (Edinburgh, 1829).

— *Biographical sketch of the late Edward Turner MD being the annual address delivered before the Harveian Society of Edinburgh on the 12th of April 1837* (Edinburgh, 1837).

Coldstream, W. (ed.), *Records of the Intelligence Department of the Government of the North-West Provinces of India during the Muting of 1857, Preserved by, and now arranged under the Superintendence of Sir William Muir, then in charge of the Intelligence Department* (Edinburgh, 1902).

Dodd, George, *The History of the Indian Revolt and of the Expeditions to Persia, China and Japan,* (London, 1859).

Donovan, Michael, 'On the Physical and medicinal qualities of Indian Hemp, cannabis indica with observations on the best mode of administration, and cases illustrative of its powers', *Dublin Journal of Science*, 26, 3 (January 1845), 368–346.

Duff, Alexander, *The Church of Scotland's India Mission. Address before the General Assembly 25 May 1835* (Edinburgh, n.d.).

Dunglison, Robley, *New Remedies: Pharmaceutically and Therapeutically Considered* (Philadelphia, PA, 1843).

Fleming, John, 'A catalogue of Indian medicinal plants and drugs, with their names in the Hindustani and Sanscrit languages', *Asiatick Researches*, 11 (1810), 153–196.

Kaye, John William, Sir, *The Life and Correspondence of Charles, Lord Metcalfe, Late Governor-General of India, Governor of Jamaica, and Governor-General of Canada; from Unpublished Letters and Journals Preserved by Himself, His Family, and His Friends* (London, 1858).

— *A History of the Sepoy War in India, 1857–1858* (London, 1864).

Latta, Thomas Aitcheson, *Letter from Dr. Latta to the Secretary of the Central*

BIBLIOGRAPHY

Board of Health, London, affording a View of the Rationale and Results of his Practice in the Treatment of Cholera by Aqueous and Saline Injections (Leith, 1832).

— 'Malignant Cholera: Documents Communicated by the Central Board of Health, London, relative to the Treatment of Cholera by the Copious Injection of Aqueous and Saline Fluids into the Veins', *The Lancet*, 18 (1832), 274–280.

Lawrie, James, 'Cases illustrating some of the Effects of Indian Hemp', *Monthly Journal of Medical Science*, 4 (1844), 939–948.

Mackintosh, John, *Principles and Practice of Physic* (London, 1836).

Meikle, George, Surgeon, EICS, 'Trial of Saline Venous Injections in Malignant Cholera: at the Drummond Street Hospital, Edinburgh', *The Lancet*, 18, 472 (1832), 748–751.

O'Shaughnessy, William Brooke, *A Manual of Chemistry arranged for Native General and Medical Students and the Subordinate Medical Departments of the Service* (Calcutta, 1842).

— 'Analysis of the Edible Moss of the Eastern Archipelago', *Journal of the Asian Society of Bengal.*

— *Bengal Pharmacopoeia and General Conspectus of Medicinal Plants* (Calcutta, 1844).

— 'Case of Tetanus, Cured by a Preparation of Hemp (the Cannabis indica.)', *Transactions of the Medical and Physical Society of Bengal*, 8 (1838–1840).

— 'Discovery of a new Principle in Human Blood in Health and Disease, and also in the Blood of Several of the Lower Mammalia', *The Lancet* (February 1835), 677–679.

— *Essays on the effects of iodine in scrofulous diseases: including an inquiry into the mode of preparing ioduretted baths/ translated from the French of M. Lugol, by W.B. O'Shaughnessy: with an appendix by the translator, containing a summary of cases treated with iodine* (London, 1831).

— 'Experimental enquiries on the laws, practical improvement and useful applications of the Galvanic Battery', *Quarterly Journal of the Calcutta Medical and Physical Society* (1 October 1837), 484–507.

— 'Experiments on the Blood in Cholera', *The Lancet* (December 1831).

— 'On the Composition and Properties of the Fucus Amylaceous', *Indian Journal of Medical Science*, 1, 1 (1834).

— 'On the Mode of Detecting Nitric Acid', *The Lancet* (May 1830), 30–33.

— 'On the Detection of Iodine, and the Hydriodate of Potash', *The Lancet* (July 1830), 633–638.

— 'On the Toxicological Relations of the Sulphocyanic Acid', *The Lancet* (2 October 1830), 33–35.

— 'On the Identity of the bark of the Strychnos Nux Vomica with the false Angustura of writers on Materia Medica', *Quarterly Journal of the Calcutta Medical and Physical Society* (1 January 1837), 9–11.

— 'Poisoned Confectionary. Detection of Gamboge, Lead, Copper, Mercury and Chromate of Lead in various articles of Sugar Confectionary', *The Lancet* (1830–1831), 193–198.

— 'Proposal of a New Method of Treating the Blue Epidemic Cholera by the Injection of highly oxygenised Salts into the Venous System, read before the

152 BIBLIOGRAPHY

Westminster Medical Society on 3 December 1831', *The Lancet*, 17, 432 (10 December 1831), 366–371.

— *Report on the Investigation of Cases of Real and Supposed Poisoning* (Calcutta, 1841).

— *Report on the Chemical Pathology of the Malignant Cholera: Containing Analyses of the Blood, Dejections, &c. of Patients Labouring Under That Disease in Newcastle and London* (London, 1832).

— 'Letter', *The Lancet* (2 June 1832).

— *On the Preparations of the Indian Hemp, or Gunjah, (cannabis Indica): Their Effects on the Animal System in Health, and Their Utility in the Treatment of Tetanus and Other Convulsive Disorders* (Calcutta, 1839).

— 'On the Existence of a New Principle (sub-rubrine,) in Human Blood in the Healthy and Diseased State, and in the Blood of Several other Mammalia', *Transactions of the Calcutta Medical and Physical Society.*

— *On the Improvement of Bengal Pottery* (s.n., 1841).

— 'On Employment of electro-magnet as moving power; with description of model machine worked by the agent', *Quarterly Journal of the Calcutta Medical and Physical Society*, 1 (1837).

— 'Memoranda relative to Experiments on the Communication of Telegraphic Signals by Induced Electricity', *Journal of the Asiatic Society of Bengal* (1840).

— 'On Narcotine as a Substitute for Quinine in Intermittent Fever', *The Lancet* (2 July 1830), 606–607, a report of his paper in the *Indian Journal of Medical Science.*

— *Selections from the Records of the Bengal Government number vii. Report on the Electric Telegraphic between Calcutta and Kedgeree* (Calcutta, 1852).

— *The Electric Telegraph in British India. A Manual of Instructions for the Subordinate Officers, Artificers, and Signallers employed in the Department* (Printed by Order of the Court of Directors, London, 1853).

O'Shaughnessy, William Brooke and Cleghorn, Hugh Francis Clarke, *Memoranda on Indian Materia Medica* (Calcutta s.n., 1838).

Pereira, Jonathan, *The Elements of Materia*, 2 vols (London, 1839).

Pettigrew, T.J., *Observations on Cholera, comprising a description of the epidemic cholera of India* (London, 1831).

Playfair, George, *The Taleef Shereef or Indian Materia Medica* (Calcutta, 1833).

Post Office Directory for Edinburgh 1828–1829, 103.

Post Office Directory for Edinburgh 1830–1831, 144.

Prout, William, 'Application of Chemistry to Physiology, Pathology and Practice', *London Medical Gazette*, 28 May 1831, 8, 257–285, 258, 268.

Ramsay, James Andrew Broun and Baird, J.G.A, *Private Letters of the Marquess of Dalhousie* (Edinburgh, London, 1910).

Rees, G.O., *On the Analysis of the Blood and Urine in Health and Disease* (London, 1836).

Reid, David Boswell, *Elements of practical chemistry. Comprising a series of experiments in every department of chemistry: with directions for performing them, and for the preparation and application of the most important tests and reagents* (Edinburgh, 1830).

Reid, Hugo, *Memoir of the late David Boswell Reid* (Edinburgh, 1863).

BIBLIOGRAPHY 153

Reynolds, J. Russell, 'On the Therapeutical Uses and Toxic Effects of Cannabis Indica', *The Lancet* (22 March 1890), 637–638.

Stevens, William, *Dr. Stevens's treatise on the cholera, extracted from his work entitled Observations on the healthy and diseased properties of the blood* (New York, 1832).

— *Observations on the Blood* (London, 1830).

— *Observations on the Nature and Treatment of the Asiatic Cholera* (London, 1853).

Squire, Peter, *Companion to the latest Edition of the British Pharmacopoeia comparing the strengths of various preparations with those of the London, Edinburgh, Dublin and United States* (London, 1884), 74–75.

Watson, Sir Thomas, *Lectures on the Principles and Practice of Physic: delivered at Kings College 1844* (Philadelphia, PA, 1858).

Wilson, Horace Hayman, *Glossary of Judicial and Rev Terms and of useful words occurring in Official Documents relating to the administration of the Government of British India* (London, 1855).

Official Documents and Publications

Official Reports made to Government by Drs Russell and Barry on the Disease called Cholera Spasmodic as Observed by them during their Mission to Russia in 1831 (London, 1832).

London Gazette, 25 October 1831, number 18864.

Papers relative to the Disease called Cholera Spasmodica in India now prevailing in the North of Europe (Printed by Authority of the Lords of His Majesty's Privy Council, London, 1831).

Rules and Regulations proposed by the Board of Health for the purpose of preventing the introduction and spreading of the Cholera Morbus. Published by a Committee of the Lords of His Majesty's Privy Council (London, 1831).

Central Board of Health, *Papers on Sanitary Instructions for Communities attacked by Spasmodic Cholera and Observations on the Nature and Treatment of the Disease drawn up by Drs Russell and Barry* (Whitehall, London, 1831).

Records of the Intelligence Department of the Government of North-Western Provinces (India) (Edinburgh, 1902).

Parliamentary Papers. House of Lords: East Indies and East India Company: reviewing his Administration in India from January 1848 to March 1856. Return To an Order of the House of Lords, dated 27th May 1856, for Copy of a Minute by the Marquis of Dalhousie, dated 28th February 1856, reviewing his Administration in India, from January 1848 to March 1856. 161. Xiii. [1]. Vol. 13.

Newspapers and Periodicals

Clendinning, John, 'Observation on the medicinal properties of *Cannabis sativa* of India', *Medico-Chirurgical Transactions*, 26 (1843), 188–221.

Lawrie, James Adair, 'Cases illustrating some of the Effects of Indian Hemp', *Monthly Journal of Medical Science*, 4 (1844), 939–948.

'Report of Royal Medico-Botanical Society, 22 February 1843', *Provincial Medical Journal and Retrospect of the Medical Sciences*, 25 February 1843, 436–438.

Robertson, William, 'Some Account of the Practice in the Cholera Hospital in Surgeon Square', *Monthly Journal of Medical Science*, 9, New Series (1849), 393–399.

Telfair, C., 'Account of the Epidemic Cholera as it Occurred at Mauritius', *Edinburgh Medical and Surgical Journal*, 17 (1821), 517–526.

Medical Reporter, Calcutta, 1895.

Secondary Sources

Ackerknecht, Erwin Heinz, *History and Geography of the Most Important Diseases* (New York and London, 1965).

Anderson, Robert D., 'The Construction of a Modern University: Age of Reform', in Robert D. Anderson, Michael Lynch and Nicholas Phillipson (eds), *The University of Edinburgh: An Illustrated History* (Edinburgh, 2003).

Anderson, Robert Geoffrey William (ed.), *Cradle of Chemistry: The Early Years of chemistry at the University of Edinburgh* (Edinburgh, 2015).

Arnold, David (ed.), *Science, Technology and Medicine in Colonial India* (Cambridge, 2000).

— (ed.), *Imperial Medicine and Indigenous Societies* (Manchester, 1988).

Arnold, Edwin, *The Marquis of Dalhousie's Administration of British India* (London, 1862).

Axelby Richard and Savitha Preetha Nair, *Science and the Changing Environment in India 1780– 1920. A Guide to Resources in the India Office Records* (London, 2010).

Bala, Poonam, *Contesting Colonial Authority: Medicine and Indigenous Responses in Nineteenth and Twentieth-Century India* (Lanham, MD, 2012).

— *Imperialism and Medicine in Bengal: A Socio-Historical Perspective* (New Delhi, 1991).

— *Medicine and Colonialism: Historical Perspectives in India* (London, 2014).

— *Medicine and Medical Policies in India: Social and Historical Perspectives* (Lanham, MD, 2007).

— 'Reconstructing Indian Medicine: the Role of Caste in Late Nineteenth-Century and Twentieth-Century India', in Poonam Bala (ed.), *Medicine and Colonialism: Historical Perspectives in India and South Africa* (London, 2014).

Bayley, Christopher Alan, *Empire and Information: Intelligence Gathering and Social Communication in India, 1780–1870* (Cambridge, 1996).

— *Imperial Meridian: The British Empire and the World, 1780–1830* (London, 1989).

— *The Raj, India and the British 1600–1847* (London, 1990).

Bhattacharya, Jayanta, 'The Genesis of Hospital Medicine in India: the Calcutta Medical College (CNC) and the Emergence of a New Medical Epistemology', *Indian Economic and Social History Review*, 51, 2 (2014), 231–264.

— 'From the Inception to the First Dissection Calcutta Medical College 1836', *Doctors' Dialogue* (2023).

BIBLIOGRAPHY

— 'From Persons to Hospital Cases: The Rise of Hospital Medicine and the Calcutta Medical College in India', *Indian Journal of History of Science* (2015).

Bidisha Chakraborty and Sarmistha De, *Calcutta in the Nineteenth Century. An Archival Exploration* (New Delhi, 2013).

Briggs, Asa, 'Cholera and Society in the Nineteenth Century', *Past & Present*, 19 (April 1961), 6–96.

Blake, P.A. 'Historical Perspectives on Pandemic Cholera', in I.K. Wachsmuth, P.A. Blake and Ø. Olsvik (eds), *Vibrio Cholerae and Cholera: Molecular to Global Perspectives* (Washington, DC, 1994).

Booth, M., *Cannabis: A History* (London, 2004).

Brahmachari, G., 'Neem-An Omnipotent Plant: A Retrospection', *ChemBioChem*, 5 (2004).

Bridge, J.A., 'Sir William Brooke O'Shaughnessy. MD, FRS, FRCS, FSA: A Biographical Appreciation by an Electrical Engineer', *Notes and Records of the Royal Society of London*, 52, 1 (1988), 103–120.

Brockington, Colin Fraser, 'Public Health at the Privy Council 1831–34', *Journal of the History of Medicine and Allied Sciences*, 61, 2 (1961), 161–195.

Bynum, W.F., *Science and the Practice of Medicine in the Nineteenth Century* (Cambridge, 1994).

Cartwright, Frederick Fox and Biddiss, Michael Denis, *Disease and History* (New York, 1972).

Choudhury, Deep Kanta Lahiri, '"Beyond the Reach of Monkeys and Men?" O'Shaughnessy and the Telegraph in India c. 1836–1856', *Indian Economic and Social History Review*, 37, 3 (2000), 331–359.

— *Telegraphic Imperialism Crisis and Panic in the Indian Empire*, c. 1830. s, Web (2010).

Christie, J.R.R., 'The Rise and Fall of Scottish Science', in M. Crosland, (ed.), *The Emergence of Science in Western Europe* (New York, 1976).

Coakley, Davis, *Irish Masters of Medicine* (Dublin, 1992).

Coley, Noel G., 'Early Blood Chemistry in Britain and France', *Clinical Chemistry*, 47, 12 (2001).

Comrie, John Dixon, *History of Scottish Medicine*, 2 vols, second edition (London, 1932).

Crawford, Dirom Grey, *Roll of the Indian Medical Service, 1615–1930*, 2 volumes (London, 1930).

— 'Notes on the Indian Medical Service', *Indian Medical Gazette*, 36 (1901), 3.

De, S.N., *Cholera: Its Pathology and Pathogenesis* (Edinburgh, 1961).

Diehl, Katharine S., 'The College of Fort William', *Libraries and Culture* (2001).

Dingwall, Helen, *A Famous and Flourishing Society: The History of the Royal College of Surgeons of Edinburgh, 1505–2005* (Edinburgh, 2005).

Douglas, Dr, 'On the use of Indian Hemp in Chorea', *Edinburgh Medical and Surgical Journal* (1869), 777–784.

Edwards, Owen Dudley, *Burke and Hare* (Edinburgh, c. 1984).

Evans, Sir Richard J., 'Epidemics and Revolutions: Cholera in Nineteenth Century Europe', *Past & Present*, 120 (1988), 121–147.

'The First Use of Intravenous Saline for the Treatment of Disease: Letter from Thomas Latta submitted to the Central Board of Health, London and published in The Lancet (1832). Preface by Jane Ferrie', *International Journal of Epidemiology*, 42, 2 (April 2013), 387–390, https://doi.org/10.1093/ije/dyt045 (accessed 8 July 2024).

Fahie, John Joseph, *A History of Wireless Telegraphy: Including Some Bare-Wire Proposals for Subaqueous Telegraphs* (Edinburgh, 1901).

Farrar, W.V., 'Science and the German University System, 1790–1850', in M. Crosland (ed.), *The Emergence of Science in Western Europe* (New York, 1976).

Gilmour, David, *The British in India. Three Centuries of Ambition and Experience* (London, 2018).

Ghosh, Suresh Chandra, *Dalhousie in India, 1848–56: A Study of his Social Policy as Governor-General* (New Delhi, 1975).

Gorman, Mel, 'Sir William O'Shaughnessy, Lord Dalhousie, and the Establishment of the Telegraph System in India', *Technology and Culture, The International Quarterly of the Society for the History of Technology*, 12, 1 (January 1971), 581–601.

— 'Sir William Brooke O'Shaughnessy FRS (1809–1889), Anglo-Indian Forensic Chemist', *Notes and Records of the Royal Society of London*, 39, 1 (1984), 51–64.

— 'Sir William Brooke O'Shaughnessy, Pioneer Chemical Educator in India', *Ambix*, xvi (1969), 107–104.

— 'Sir William O'Shaughnessy, Introduction of Western Science into Colonial India: Role of the Calcutta Medical College', *Proceedings of the American Philosophical Society*, 132, 3 (1988).

Grant, Alexander, *Physician and Friend. His Autobiography and his Letters from the Marquis of Dalhousie*, edited by George Smith (London, 1902).

Greig, E.D.W., 'The Treatment of Cholera by intravenous Saline Injections: With Particular Reference to the Contributions of Dr Thomas Aitcheson Latta of Leith (1832)', *Edinburgh Medical Journal*, 53 (1946), 256–263.

Guy, Geoffrey W., Whittle, Brian A., and Robson, Phillip J. (eds), *The Medicinal Uses of Cannabis and Cannabinoids* (London, 2004).

Hamlin, Christopher, *Cholera: The Biography* (Oxford, 2009).

Hamilton, David, *The Healers. A History of Medicine in Scotland* (Edinburgh, 1981).

Harland-Jacobs, Jessica L., *Builders of Empire. Freemasons and British Imperialism, 1717–1927* (Chapel Hill, NC, 2013).

Harris, Christina Phelps Harris, 'The Persian Gulf Submarine Telegraph of 1864', *Geographical Journal*, 135, 2 (1969), 169–190.

Harrison, M., 'Disease and Medicine in the Armies of British India, 1750–1830: The Treatment of Fevers and the Emergence of Tropical Therapeutics', *Clio Med.* 8 (2007), 87–119.

Howard-Jones, Norman, 'Cholera Therapy in the Nineteenth Century', *Journal of the History of Medicine* (1972), 373–395.

Kejariwal, O.P., *The Asiatic Society of Bengal and the Discovery of India's Past 1784–1838* (New Delhi and Oxford, 1999).

BIBLIOGRAPHY

Kieve, Jeffrey, *The Electric Telegraph. A Social and Economic History* (Newton Abbot, 1973).

Kumar, Deepak (ed.), *Disease and Medicine in India. A Historical Overview* (New Delhi, 2001).

Kumar, Deepak and Macleod, Roy (eds), *Technology and the Raj: Western Technology and Technical Transfers to India, 1700–1947* (New Delhi, 1995).

Kumar, Deepak and Raj Sekhar Basu (eds), *Medical Encounters in British India* (New Delhi, 2013).

Lee-Warner, Sir William, *The Life of the Marquis of Dalhousie KT* (London, 1904).

Leslie, Charles (ed.), *Asian Medical Systems: A Comparative Study* (Berkeley, CA, 1976).

MacDonald, Donald, 'The Indian Medical Service. A Short Account of its Achievements 1600–1947', *Proceedings of the Royal Society of Medicine*, 49 (1956), 13–17.

MacGillivray, Neil, 'The Congested Districts Board and the Isleornsay Pier, Isle of Skye, 1899–1906', *Northern Scotland*, 13, 1 (May 2022), 45–62.

— 'Sir William Brooke O'Shaughnessy (1808–1889), MD, FRS, LRCS Ed: Chemical Pathologist, Pharmacologist and Pioneer in Electric Telegraphy', *Journal of Medical Biography*, 25, 3 (2015), 186–196.

McGrew, Roderick E., 'The First Russian Cholera Epidemic: Themes and Opportunities', *Bulletin of the History of Medicine*, 36 (1962), 220–244.

Makepeace, Margaret, *East India Company London Workers. Management of the Warehouse Labourers 1800–1858* (Woodbridge, 2010).

Masson, A.H.B., 'Latta: Pioneer in Saline Infusion', *British Journal of Anaesthesia*, 43 (1971), 681–685.

Mills, Jams H., 'Colonising Cannabis: Medication, Taxation, Intoxication and Oblivion, c. 1839–1955', in R. Deb Roy and G.N.A. Attewell (eds), *Locating the Medical: Explorations in South Asian History* (New Delhi and Oxford, 2018).

— *Cannabis Britannica: Empire, Trade, and Prohibition, 1800–1928* (Oxford, 2002).

Morrell, Jack B.,' Practical Chemistry in the University of Edinburgh, 1799–1843', *Ambix* , 16 (July 1969), 68–80.

Morris, Robert John, *Cholera 1832. The Social Response to an Epidemic* (London, 1976).

Morus, Iwan Rhys, 'The Electric Aerial: Telegraphy and Commercial Culture in Early Victorian England', *Victorian Studies*, 39, 3 (1996), 339–378.

Nalin, David R., 'The History of Intravenous and Oral Rehydration and Maintenance Therapy of Cholera and Non-Cholera Dehydrating Diarrhoeas: A Deconstruction of Translational Medicine: From Bench to Bedside?', *Tropical Medicine and Infectious Disease*, 7.50 (2022), 1–28.

Patil, Aishan and Leach, J.P., 'Latta: "Physician Worth His Salt": An Eclectic View of Thomas Aitchison Latta's True Contribution to Modern Medicine', *International Journal of Head Neck Surgery*, 13, 3 (2021), 92–97.

Pelling, Margaret, *Cholera, Fever and English Medicine 1825–1865* (Oxford, 1978).

Peterson, M. Jeanne, *The Medical Profession in Mid Victorian London* (Berkeley, CA, 1978).

Pisanti, Simona and Bifulco, Maurizio, 'Medical Cannabis: A Plurimillennial History of an Evergreen', *Journal of Cellular Physiology* (2019), 8342–8351.

Pollitzer, Robert, *Cholera* (Geneva, 1959).

Porter, Roy, *The Greatest Benefit to Mankind. A Medical History of Humanity from Antiquity to the Present* (London, 1999).

Proudfoot, Alex T., Good, Alison M. and Bateman, D. Nicholas, 'Clinical Toxicology in Edinburgh, Two Centuries of Progress', *Clinical Toxicology* (2013), 509–514.

Risse, Gunter B., *New Medical Challenges during the Scottish Enlightenment* (Amsterdam and New York, 2005).

Roy, Kaushik, *War and Society in Colonial India, 1807–1945* (Oxford, 2006).

Nenadic, Stana, 'Writing Medical Lives: Creating Posthumous Reputations: Dr Matthew Baillie and his Family in the Nineteenth Century', *Social History of Medicine*, 23, 3 (2010), 509–527.

Shridharani, Krishnalal, *Story of the Indian Telegraphs. A Century of Progress* (New Delhi, 1954).

Sirkin, Natalie R. and Sirkin, G., 'The Battle of Indian Education. Macaulay's Opening Salvo Newly Discovered', *Victorian Studies* (1971), 407–428.

Sutton, Jean, The East India Company's Maritime Service 1746–1834: Masters of the Eastern Seas (Woodbridge, 2010).

Thomas, D.P., 'The Demise of Bloodletting', *Journal of the Royal College of Physicians of Edinburgh*, 44 (2014), 72–77.

Thomas, G., 'O'Shaughnessy's Experiments in Colour Photography', *History of Photography*, 10, 2 (1986).

Underwood, E. Ashworth, 'The History of Cholera in Great Britain', *Proceedings of the Royal Society of Medicine*, 41, 3 (1948), 165–173.

Venters, George, 'Leith in the Time of Cholera- the Story of Thomas Latta', *Hektoen International. A Journal of Medical Humanities*, 7, 1 (2015).

Wenzlhuemer, Roland, *Connecting the Nineteenth-Century World: The Telegraph and Globalization* (Cambridge, 2013).

Whitfield, Michael J., 'Dr John Tytler (1787–1837), Superintendent of the Native Medical Institution, Calcutta', *Journal of Medical Biography*, 28, 4 (2021), 184–189.

Wusjastyk, Dominik, 'Medicine, India', in Maryanne Cline Horowitz (editor in chief), *New Dictionary of the History of Ideas* (Farmington Hills, MI, 2005), 1411–1413.

Journals and Newspapers

Ambix
Asiatic Journal and Monthly Register for British India
Bombay Gazette
British Journal of Anaesthesia
British Medical Journal

BIBLIOGRAPHY   159

Caledonian Mercury
Calcutta Review
Calcutta Monthly Journal
Clinical Toxicology
Clinical Chemistry
Clio Med.
Dublin Journal of Medical Science
East India and Colonial Magazine
Edinburgh Medical and Surgical Journal
Friend of India
Fife Herald
Geographical Journal
Glasgow Medical Journal
Hull Advertiser
Indian Economic and Social History Review
Indian Journal of the History of Science
The International Quarterly of the Society for the History of Technology
Journal of the Asiatic Society of Bengal
Journal of the History of Medicine and Allied Sciences
Journal of the Kilkenny and South-East of Ireland Archaeological Society
Journal of Medical Biography
Journal of Postgraduate Medicine
Journal of the Royal College of Physicians of Edinburgh
Journal of Cellular Physiology
Journal or Transactions of the Medical and Physical Society of Calcutta
The Lancet
London Medical Gazette
London Medical and Physical Journal
Macmillan's Magazine
Monthly Journal of Medical Science
Morning Herald
Past & Present
Proceedings of the American Philosophical Society
Proceedings of the Royal Society of Medicine
Provincial Medical Journal and Retrospect of the Medical Sciences
Quarterly Journal of the Medical and Physical Society of Calcutta
Resuscitation
The Times
The Scotsman
Social History of Medicine
Victorian Studies

Unpublished Theses

Bala, Poonam, 'State and Indigenous Medicine in Nineteenth- and Twentieth-Century Bengal: 1800–1947, Unpublished PhD thesis, University of Edinburgh, 1987.

Christison, Alexander, 'Cannabis Indica', Unpublished doctoral thesis, University of Edinburgh, 1850.

Latta, Thomas, 'De Scorbuto', Unpublished MD thesis, University of Edinburgh, 1919.

MacGillivray, Neil, 'Food, Poverty and Epidemic Disease. Edinburgh 1840–1850', Unpublished doctoral thesis, University of Edinburgh, 2004.

O'Shaughnessy, William Brooke, 'De Metastasi Rheumatism Acuti', Unpublished MD thesis, University of Edinburgh, 1829.

INDEX

Adamson, Dr John
  photography pioneer 94 n.16
Adley, Charles·C
  East India Railway Company civil
      engineer 118–119
Agra
  1857 insurrection: lines destroyed
      from Agra to Indore, Delhi and
      Cawnpore 133
  centre of communications 126
  first telegraph message to Calcutta
      March 1854 127
Ainslie, Dr Whitelaw
  author of *Materia Medica of
      Hindoostan* 62–64 n.21
  *Bengal Dispensatory and
      Pharmacopoeia* 88 n.1
Alexander, William 107 *see also* Kemp
  practical scheme for
      telegraphy 110–111
Alison, Professor William Pulteney 2, 12
  O'Shaughnessy clinical assistant
      to 16, 17 n.20, 24
  social conditions of the poor 141
Annesley, Dr James
  on bleeding 28, 29 n.10
Arnold, David 67 n.28
Arsenic
  as chemical examiner 35, 39
Asiatic Society
  foreign secretary to Medical and
      Physical Society 69–71
  Sir Charles Metcalf and
      O'Shaughnessy members of
      Asiatic Society 90
  O'Shaughnessy, secretary of Asiatic
      Society 7
  papers on Cannabis 58, 63 n.18
  photographic drawings shown to
      Society 1839 94
  promotion of telegraphy 114

published research on Cannabis
    Indica 75–76
H.H. Wilson as secretary 98
Assay Master 87, 106, 116
Auckland, Lord 104
Ayurveda 5, 59, 62, 66, 70, 88, 142

Babbage, Charles 92
Baily, Francis 93
Bala, Professor Poonam 59 n.6, 67 n.26
Banks, Sir Joseph 63
Barker, Professor Francis
  Professor of Chemistry, Trinity
      College, Dublin 11
Barry, Sir David 33–34, 50
Bazin, Adolphe, Baron du Chonay 114,
    115 n.29
Behar 69
Bell, Dr George Hamilton 27, 28 n.7,
    54
Bentham, Jeremy 26
Bengal
  Anglicists versus Orientalists 88–90
  dispensatory 57–58, 60
  O'Shaughnessy as chemical
      examiner 116
  O'Shaughnessy monograph on electric
      telegraphy 115
  promoter of education of the natives of
      Bengal 97–98
  telegraphy in un-surveyed Bengal 107
*Bengal Dispensatory and
    Pharmacopoeia* 4
  deaths of colleagues from disease 138
  five sacred herbs 142
  reference to works of Pereira, Royle
      and Lindley 102
  single handed editor 87–88
  students at the Medical College 61
Bengal Medical Service 1, 3, 5, 91, 138,
    147

162 INDEX

Bengal Mint 4, 87, 96
 Nundy as assistant in 130
 O'Shaughnessy as deputy assay master
  1844 102, 116–118
Bentinck, Lord William, 60 n.8, 61
Bhang 62, 142
Bleeding 27–29, 36, 49, 139–141
Bombay 102–103, 106, 117, 127, 134
Bombay Gazette 95
Bombay Times 94
Boswell, Sarah 8
Botany 12, 48, 99, 100
Brendish, William 131
Bridge, J.A. 1
British and Foreign Medical Review 99
British Medical Journal 137
Brooke, Henrietta 8
Brooke, Henry 8
Brooke, Sir William O'Shaughnessy 148
Brooke, William Joseph Eyre,
 Lieutenant-General 8
Burke and Hare 15, 45
Bynum, William F. 53–54

Calcutta 5, 104, 106, 144–145
 Botanic Garden 48, 63, 105, 113
 Calcutta Review 119, 122, 125–126
 end of O'Shaughnessy's tenure of
  chair in Calcutta Medical
  College 89
 experimental line Calcutta to Diamond
  Harbour 1851–1852 121
 Medical College 4–5, 61 n.12, 62,
  69, 87
 mint 60, 106, 116, 129
Caledonian Mercury 110, 141 *see also*
 Alexander; Kemp
Campbell, Sir Colin 129, 134
Canada 4, 90, 102
Cannabis Indica 7–8, 61, 62 n.13, 62
 n.16, 63
 *British and Foreign Medical
  Review* of O'Shaughnessy on
  Cannabis 77
 clinical trials 88
 Grimson 142–144
 lecture on cannabis to Royal Botanical
  Society, London 89
 medical use in USA 85
 Pandit Gupta 76
 Pereira and cannabis 100

physicians in Britain on use
 of 80–83
Playfair description in *Taleef
 Shareef* 75
Cawnpore 134
Central Board of Health 25–26, 32, 34,
 37–38, 43 n.43, 43 n.44
 Latta's letter to Board 9, 41–42, 42
  n.42
 Lewin's accusations 52
 O'Shaughnessy's
  recommendations 140
 Sir James McGrigor 100–101
Chadwick, Sir Edwin 141
Chemical Examiner 116 *see also* Bengal
Chemistry 2–4, 24–25, 33
 chemistry Dublin and
  Edinburgh 11–12, 13n10, 14 n.13
 forensic chemical analysis 58
 Herschel and chemistry 93
 Kemp on chemistry 108–109 *see also*
  Kemp
 O'Shaughnessy appointed Professor
  of Chemistry Calcutta Medical
  College 69
 practical chemistry 36
 Professor of Chemistry, Calcutta 142
 Royal Society citation 92
 scientific method 66
 studies in Edinburgh 116
 text on chemistry 87
Cholera 20–21, 23 n.1, 25, 26 n.4, 27
 n.5, 27 n.6, 28 n.7
 address to Westminster Medical
  Society 35, 36 n.29
 chemistry in Edinburgh 36
 cholera in Bengal 88
 death of a young woman from
  cholera 34
 Professor Robert Graham's letter to Dr
  Nathaniel Wallich 48 *see also*
  Wallich
 Herschel to Francis Baily 93
 history of intravenous and oral
  rehydration 40–41, 41 n.41
 humoral theory 138–139
 Latta's letter to Board of Health 42
 Latta's technique 47
 Sir James McGrigor's and
  O'Shaughnessy's cholera blood
  analysis 100

INDEX

observations on cholera in
  Sunderland   32 n.18, 32 n.19
quarantine   29–31
Royal Society citation   91
Dr Stevens on cholera   33 *see also*
  Stevens
statement to Board of Health   37–38
Chorea   81
Christie, Dr Alexander Turnbull   29 n.1,
  30–31
Christison, Alexander, later Sir Alexander
  Bart   80 *see also* cannabis
Christison, Robert, later Sir Robert
  Christison Bart, Professor of
  Forensic Medicine   12, 14, 17 n.20,
  18
  advice to Dutch Government on saline
    treatment   52, 52 n.62
  Edinburgh Board of Health   140
  interviewed Knox   16
  London Medical Gazette   140 n.7
  toxicology in Edinburgh   39 n.34, 39
    n.35, 36, 38
  use of cannabis therapeutically   80
Clarke, Sir James   98–99
Cohen, Professor Samuel K Jr   45 n.47
Constantinople   145
Cooke, Sir William Fothergill   106–108,
  111
County Clare   68
Coupar, Sir George, Bart   106
  letter from Dalhousie   128
Craigie, Dr Thomas   49
Cullen, William   139–140

Daily News   131
Daguerrotype   93–94
Dalhousie, Lord   4, 103–104, 106, 110,
  116–120 *see also* electric telegraph
  and telegraphy
  Bombay government telegraph
    message   133
  Dalhousie as promoter of
    telegraphy   144
  Dalhousie sends O'Shaughnessy
    to meet with Court of
    Directors   125–128
  Dalhousie's policy on Indian
    kingdoms   131
  doctrine of Lapse   117
  Minute of 1852   124

Darwin, Charles   15 n.15
Delhi   131–133
Diamond Harbour   121–124, 130, 144
Dixon, Professor W.E.   83
Donovan, Dr Michael   79 n.55
Douglas, Dr Andrew Halliday   81 n.60
Drummond Street, Edinburgh   44–45,
  47–48, 55
Dublin   7–8, 35
  letter to H.H. Wilson,1843   90
Duff, Revd Alexander
  Anglicist   61
  errors in Hindu learning   67 n.9
Dum-Dum   133
Dunglison, Professor Robley   79

East India Company   3, 5, 21, 48, 58
  Adley and telegraph proposal   118
  Anglicist-Orientalist debate   60
  cannabis   75, *see also* Cannabis
  Court of Directors approve extension
    of telegraph line   126, 130
  decision to establish electric telegraph
    in India   103
  election to Royal Society   96–98
  financial benefits   83
  Fort William College and Indian
    languages   68, 71
  Gilchrist and Urdu Dictionary   68
  line through Persian Gulf   145
  malaria and its treatment   72–74
  medical furlough   87–89
  new medicaments   67
  Scottish surgeons   62–66
  telegraph line financed by
    Company   111
Edinburgh   2 *see also* chemistry and
  telegraphy
  Board of Health opposition to
    intravenous saline   140
  Botanic Garden, now Royal Botanic
    Garden   48 n.53, 80
  Edinburgh New Philosophical Journal
    109
  Medical School   7, 11, 24
  New Town Dispensary   17
  scientific method and the Scottish
    Enlightenment   66
Electric telegraph   103–112 *see also*
  telegraphy
  arrival of Dalhousie   116–117

Electric telegraph (*continued*)
Dodd, author of *History of Indian Revolt* on electric telegraph 128
experiments with underwater telegraphy 144
first Line in India opened 1852 124 n.52, 53
Nundy, Seebchunder 130–131
O'Shaughnessy sent to London to Court of Directors 126
reference to Volta and Galvani 114
telegraph role in Uprising 1857 132–138
Ennis *see also* Smith, Erasmus
Grammar School 9–11, 24

Fleming, Dr John 63, 64 n.22, 65 n.23
Forbes, Sir John 98
Forbes, Major-General William Nairn 119 n.41, 120
Forensic medicine 12, 16–17, 38, 58, 138
Fort William 68 n.36, 71
Freemasonry 97 n.22, 97 n.23, 123 n.50, 143

Gaelic 96
Galvanism 90, 93, 108–109, 112–113
Ghose,Saroj 11 n.21, 133 n.72
Gilchrist, John Borthwick 68 n.30
Girdwood, Dr G.F., 50, 50 n.58
Glendinning, Dr John 78
Goodeve, Professor Henry Hurry 76
Gorman, Professor Mel 1 n.2, 12, 18 n.11
on O'Shaughnessy and telegraphy 121, 123
on role of Dalhousie in promotion of telegraphy 116
Graham, Professor Robert 12, 48 *see also* botany; Edinburgh
Grant, Sir John Peter 97, 116, 123 n.50
Grant and freemasonry 144
Graves, Dr Robert 79
Gregory, Professor James 139
Gregory, James Crauford 18, 19, 36 n.31
return to Edinburgh, new edition of Cullen 52, 53 n.63, 54
son of Professor James Gregory 139
Grimson, Dr Lester 142
Gunja 57, 62 n.13

supplied to Mr Squire 77
Gupta, Pandit Madhusudan 73 n.43, 76

Henry, Professor Joseph 4, 90, 111–112, 121, 144
Herschel, Sir John Bart 92–96
History of Photography 94 *see also* Adamson
Hooghly River 105, 121, 123–124
disposal of wrecks 95
underwater telegraphy 144
Hope, Professor John 64 *see also* botany; Edinburgh
Hope, Professor Thomas 11, 13, 24, 64, 66, 116
Hope at demonstration of electric telegraph in Edinburgh 108 *see also* Kemp
Horsfield, Dr Thomas 101
Howard-Jones, Norman 7, 41
Hume, Joseph MP 3, 7, 19 n.25, 21, 83, 104 n.3
humoral theory 7, 33, 49, 51, 138–139
hydrophobia 76
hypodermic syringe 84

insurrection of 1857
Sir Colin Campbell 134
imperial narrative 143, 145
influence of telegraph 103
Nundy's service 130–131

Jervis, Thomas Best, Lieutenant-Colonel 101
Journal of the Asiatic Society of Bengal 115

Keats, John 99
Kedgeree 123–124, 127, 114
Kell, Dr James Butler 32
Kemp, Kenneth 108–111 *see also* Alexander
Khaleeli, Zhaleh 67 n.27
Kiev, Jeffrey 110
Kings College, London 99, 138
Knighthood 128, 138
Knox, Dr Robert 12, 15–16
Koch, Robert 141

Lahore 103, 124, 127
extension of telegraph to 129–132

INDEX

Lancet 2–3, 7
  1854 *Lancet* review of telegraphy 145
  assistant to Professor W.P. Alison 24
  Dr George Hamilton Bell on
      venesection 54
  Dr A. Brierre de Boismont on blood in
      cholera 27
  Latta on O'Shaughnessy's report 42
  Latta on saline and support of Dr
      Lewins 46–49
  Dr George Meikle on saline 45n47
  monograph on treatment of
      scrofula 18 n.24
  O'Shaughnessy's first
      publication 16–17
  opposition to saline in Edinburgh 140
  Dr J. Russell Reynolds on
      cannabis 81
  speech as secretary of group
      promoting London College of
      Medicine 19 n.25
  Sunderland and Newcastle 33–37
  tetanus, a new remedy 59 n.4
  use of narcotine in fever 73 n.44, 74
Latta, Dr Thomas Aitcheson 23, 28
  death of Latta 54–55
  intervention of Dr James Craufurd
      Gregory 140
  intravenous saline 100
  Dr James Macintosh, 141
  saline first used 40–50
  saline treatment 2–7
  saline use in Bengal 142
Lawrie, Dr James Adair 81
Lawrence, Helen and Isabella 16
Lawrence, Sir John, Chief Commissioner
      Punjab 132
Leith 46–47, 52 *see also* Latta
Lewins , Dr Robert 41, 44, 46, 49, 52,
      140 *see also* Latta; Leith
Liebig, Professor Justus von 36
Limerick 7, 9, 137
Lindley, Professor John 88 n.1, 100–102
London College of Medicine 3, 19,
      20 n.26, 21, 34 *see also Lancet*;
      Wakley
London Electrical Society 90, 113
Lucknow 133 *see also* Campbell;
      Stewart
Lugol, Dr Jean 25

Macaulay, Thomas Babington 5 n.4,
      60–61, 88 n.9, 90
Mackintosh, Dr John 44 n.45, 45, 47–49,
      49 n.55, 51 n.60, 55, 141
Maclean, Dr William 52, 53 n.63
Macleod of Geanies, Donald 139
Macmillan's Magazine 131 *see also*
      Brendish; insurrection; telegraph
McCabe, Dr James 50, 51
Madras 103, 117, 124, 127–128
Malaria 72, 88
McMahon, Sir John, Bart 8
Martin, Dr Robert Montgomery 61
Masson, Dr Alastair H.B. 6
Mead, Alice 84 n.68
Medical Gazette 100
Medical Reporter 145
Medical and Physical Society of
      Calcutta 63, 65, 72–73, 98
  O'Shaughnessy appointed foreign
      secretary 69 n.32
  O'Shaughnessy on galvanic
      batteries 109
Meerut 133
Meikle, Dr George 44, 45 n.47, 51
Metcalfe, Sir Charles 4, 70, 90 n.6, 91,
      97, 101
Mikuriya, Dr Tod Hiro 84
Mill, James 5, 19 n.25
Mills, Professor James H. 83
Montgomery, Sir Peter 132
Morning Herald 95
Morse, Samuel Finlay Breese 4, 90,
      111–112, 128
Munro, Alexander Tertius 15

Nalin, Professor David 33 n.33, 40 n.38
  oral replacement therapy 49, 50 n.56
narcotine 57, 73 n.44, 74
Native Medical Institution 61 *see also*
      Tytler
neem 72 n.41
Nehru, Jawaharlal 105
Newcastle 2, 3, 23, 30–31, 37–38
Newhaven 46 *see also* Latta
Noltie, Dr Henry 64 n.22, 74 n.47
Nundy, Seebchunder
  assistant to O'Shaughnessy at Calcutta
      Mint 106
  telegraph pioneer and sender of
      first signal from Diamond
      Harbour 130

obituary in *British Medical Journal* 137
opium 69, 73, 77–80, 83–84
Orfila, Dr Mathieu 36–39
O'Shaughnessy, Captain Daniel 8–9
O'Shaughnessy, Revd Dr James, Bishop
    of Killaloe 10
O'Shaughnessy, Richard 76
Ootacamund 125, 127
Oxford University 20, 96–97

Palmerston, Lord 145, 147
Pasley, General Sir Charles
    William 94–96
Pereira, Dr Jonathan 82, 88 n.1, 90, 100,
    102
Peshawar 132
Philosophical Radicals 19
photography 6, 87–88, 93–94, 102
Playfair, Dr George 63–66, 75
Playfair, Lyon, later Lord Playfair 65
Pollitzer, Robert 142
Prinsep, James 63
Prout, William 24, 25 n.3

quarantine 29–30, 33 *see also* cholera
quinine 72–74, 88 *see also* malaria;
    opium

Rangoon, capture of 144
Reid, Dr David Boswell 13, 14 n.13, 17,
    24
Reynolds, Dr J Russell 78, 81–82 *see
    also* Squire; Pereira
rohema 72
Rolleston Committee, 1924 83
Royal Asiatic Society 89, 98
Royal Botanical Society 89
Royal College of Physicians of
    Edinburgh 7, 48–49, 54, 139
Royal College of Physicians London 18
Royal College of Surgeons of
    Edinburgh 13 n.12, 14, 16, 18
Royal Medico-Botanical Society 82
Royal Society 4, 89,
    citation for fellowship 91, 92–98,
        101–102, 116, 137, 143
Royal Statistical Society 98
Round down Cliff 95 *see also* Royal
    Society; Pasley
Roxburgh, Dr William 64 n.19, 20, 88
    n.1
Royle, John Forbes 88 n.1, 99, 102

Russell, Sir William 34
Russia 25–26, 38

Sands, Daniel 8
Sabine, Julia Greenly 148
saline 2, 6, 24, 42 n.42
    cure rate at Drummond Street
        Hospital 44
    Gregory on intravenous
        saline 140–141
    McCabe on humoral pathology 51
    reports from physicians 50
    use of at Drummond Street cholera
        hospital 48–49
Sanskrit 96
scientific method 58, 66
Scotsman 111, 141
Shridharani, Krishnalal 105, 129, 134
Singh, Kumar 52 n.6
Smith, Erasmus 9 n.5, 10 n.6, 11 *see also*
    Ennis
Smithsonian Institute 112
Society of Apothecaries 20
Sproat, William 31
Squire, Peter 77, 78 n.54, 81*see also*
    Donovan; Douglas; Dungliston;
    Graves
Stevens, Dr William 33 n.21, 37, 40
Stewart, Lieutenant-Colonel Patrick
with Sir Colin Campbell 134
    extension of line to Lahore 129
    telegraph line to Persian Gulf 147
Sunderland 2–3, 21, 24–25, 29–31, 32
    n.18, 19, 33 *see also* Sproat
    intimation of intravenous
        treatment 35 n.27
    O'Shaughnessy description of the
        death of a young woman 34,
        34 n.24
    O'Shaughnessy's findings
        communicated to Board of
        Health 37
Sykes, Colonel William Henry 98

Taleef Shareef 65–66
Talbot, William Fox 103
Telegraphy 1, 4, 6, 87–88, 101–103
    experimental line in Edinburgh 108
    first line in England 107
    Kemp's original work 109, 112, 129,
        143 *see also* electric telegraph
tetanus 59 n.4, 76, 82, 14

INDEX

The Times   30, 103, 112, 124,128,131
Todd, Charles   132
Trevelyan, Charles E, later Sir Charles
   Bart   5, 60–61, 70 n.37
Trinity College, Dublin   2 n.6, 11 n.7, 25,
   29 n.10
Turner, Edward   13, 14 n.13, 17, 24
Tytler, Dr John   61, 97

UK Pharmacopoeia   83
Unani   5, 59, 62, 66, 72
Underwater explosions   95–96
USA Pharmacopoeia   83

Wakley, Thomas   3–4, 7, 9, 19–21, 25
   n.24, 26–29, 34, 47
Wallich, Dr Nathaniel   48, 70, 88 n.1,
   105 n.6,113–114
Watson, Sir Thomas   40 n.39, 41
Wheatstone, Sir Charles   102, 106–111,
   115
Wilson, Professor Horace Hayman   4,
   60, 70 n.37, 89–90, 96–98, 102,
   144
Wright, Dr J.   74–75
Wusjastyk, Dr Dominik   59